Rushall

Rushall

The Story of an Organic Farm

Barry Wookey
with a foreword by Lord Penney, OM, FRS

Photographs by Rosemary and Penelope Ellis

Basil Blackwell

First published 1987
Reprinted, updated and first published in paperback 1988.

Basil Blackwell Ltd
108 Cowley Road, Oxford, OX4 1JF, UK

Basil Blackwell Inc.
432 Park Avenue South, Suite 1503
New York, NY 10016, USA

British Library Cataloguing in Publication Data

Wookey, Barry
    Rushall : the story of an organic farm.
    1. Organic farming — England
    I. Title
    631.5'84'0942   S605.5

    ISBN 0-631-15526-0
    ISBN 0-631-16308-5 Pbk

Library of Congress Cataloging in Publication Data

Wookey, Barry, 1925 —
    Rushall, the story of an organic farm.

    Includes index.
    1. Organic farming — England — Wiltshire.
2. Rushall Farm (England)   I. Title.
S605.5.W586   1987   631.5'84   86-32726
ISBN 0-631-15526-0
ISBN 0-631-16308-5 (pbk.)

Typeset in 11 on 13 pt Atlantic
by Pioneer Associates, Perthshire
Printed in Great Britain by T. J. Press (Padstow) Ltd, Padstow, Cornwall

of nutrients. Algae and other water vegetation proliferate, oxygen is depleted and some fish populations are eliminated. Eutrophication, as this process is called, is, however, a complex phenomenon. Sometimes the main cause is fertilizers. Animal manures freshly washed by rain can contribute. In other cases, sewage not fully treated is, or has been, a major source. Phosphates from detergents now make only a minor contribution. The eutrophication of Lake Erie and other lakes produced widespread concern in the early seventies. That such a large body of water could be affected shocked not only environmentalists but also many others.

In 1970 the World Health Organization (WHO) made recommendations about the purity of drinking water and set a maximum acceptable level of nitrate concentration for European conditions. The British water authorities adopted these recommendations, which were confirmed by the WHO in 1977. In 1980 the EEC set the maximum acceptable normal level of nitrate concentration at half the WHO limit, although derogations were allowed at the discretion of national authorities. In the United Kingdom, the EEC limit now normally applies, but derogations to higher levels approaching the WHO limit have been authorized as well as three-month average levels at up to 80 per cent of the WHO limit. There can be no certainty that these derogations will be allowed indefinitely, because the current tendency in the West is for public health limits to be tightened.

The area supplied by the Anglia Water Authority appears to have the highest general levels of nitrate in the United Kingdom. The concentration in some rivers exceeds the EEC limit for one to three months in the autumn or winter. Some of the aquifers beneath arable land have nitrate levels exceeding the EEC limit and in a few cases exceeding even the WHO limit. Levels in the aquifers are rising at about 1 per cent of the WHO limit per year and mathematical models predict an ultimate rise to twice the WHO limit if agricultural practices continue unchanged. Other water authorities with large areas of high-productivity arable farming are now finding nitrate levels in some areas not very different from those in Anglia.

Water authorities may soon have to face considerable expenditure in order to keep the blends of water coming from the tap within legal limits. Whichever technical options are chosen to reduce nitrate levels, including several palliative measures, the costs must be regarded as subsidizing the use of nitrates on farms. If some were to change to organic methods, less nitrate would be used and the cost of control would be less.

Many other countries are at about the same stage of questioning as the United Kingdom. Responsible committees are evaluating national positions and before long there will be many statements and proposals for action.

Within the EEC (though similar considerations apply elsewhere) there are three principal reasons for encouraging the adoption of organic methods. First, many people today are deeply concerned about the speed at which man is changing the natural world without fully understanding the consequences. The quality of the environment and the well-being of future generations of many forms of life may be at risk. For example, industrial pollution and the destruction of forests have damaging effects on the climate, the land and the whole ecological system, threatening the extinction of species of wildlife and placing food and water supplies at risk. There are similar dangers with farming methods dependent on the heavy use of chemicals. Organic farming, which rejects such methods, fits in well with the general philosophy of safeguarding the natural world instead of exploiting it with little regard for the consequences.

Second, changing at least some farms to organic methods would help reduce the concentration of agricultural chemicals in rivers, lakes and underground waters, so reducing the potentially enormous cost of treating water supplies to make them safe for consumption.

Third, changing an appreciable number of farms to organic methods would help reduce the Community food mountains and increase the production of high-quality, healthy food — food that many people prefer and for which at present they are prepared to pay a premium. It is worth emphasizing that there would be no loss of farm land and no loss of jobs.

Organic farming in the United Kingdom now accounts for barely 1 per cent of total food production. This is not enough to have any effect on food mountains or nitrate concentrations. If 10 per cent of food were produced by organic methods, then the effect would be appreciated and agriculture would have a strategic reserve of land, crops and animals which would not rely on the scientist to protect them from nature. However, the reduction in cereal production would amount to less than half the recent variations between good and poor seasons and therefore in economic terms would not go far enough. There would have to be still greater use of organic methods or some reduction in food production on conventional farms, and with either there would be a further reduction in the use of chemicals.

The Community will have to do something soon about the food mountains, first to stop the flow of food into the mountains from increasing, and then to reduce the size of the mountains to reasonable stockpile reserves. The politicians have the difficult duty on behalf of us all to avoid using Community money to produce food mountains while at the same time ensuring that farms and farmers have a good future. Assuming that the politicians can agree on modifications to the Common Agricultural Policy, some of the money saved will be needed to persuade farmers to co-operate in the new policy. Farmers who are prepared to change to organic methods and persist with them should be assisted financially. The inducements must be generous. Otherwise, we shall get short-term solutions which later on we may regret.

One of the best of Britain's organic farmers is Barry Wookey. He is known internationally and he has many people visit his farm to see and hear at first-hand what he is doing. During the last fifteen years he has put his heart and soul into organic farming and has gradually converted his 1650-acre Wiltshire farm from conventional to wholly organic methods. Nobody could speak with greater authority about the practical problems of running a profitable organic farm, including keeping it solvent during the early stages, or about the satisfaction that running such a farm can give. His book is a testament of faith by a farmer who cares deeply about the land and its creatures and plants. His arguments and his experiences deserve the most serious attention.

Preface

This book has been written in an attempt to help the many people who are becoming increasingly worried by the trends of modern agriculture. The general public is worried because of the danger of residues in the food they eat, and farmers are worried at the rapid fall in profitability they are experiencing.

We have many visitors to Rushall, both professional farmers and interested observers, and the two remarks we most frequently hear after they have been round the farm are 'Amazing — I wouldn't have thought it possible', and 'I would like to do something like this if only I could afford to.'

It is hoped that this book will do something for both sets of people — encourage the farmers who are thinking of converting their farms to an organic system and demonstrate to other concerned people that there is an active core of organic farmers in this country who are trying to find a viable system that does not rely on huge chemical inputs and that can be sustained for the rest of this century and beyond.

1
Rushall

Rushall Farm is situated in and around the village of Rushall, which lies on the River Avon at the southern end of Pewsey Vale and close to the northern edge of Salisbury Plain. The village consists of about forty dwellings, from the sixteenth-century thatched cottages to a few modern brick houses. It is fortunate in having both a church and a school, and, although it has never boasted a pub or a village shop, it did until recently have a post office and a blacksmith.

The church, St Matthew's, stands on the site of an earlier building that was standing when the Domesday Book was compiled, in 1086. The present church was built in 1332, of brick, stone and flint and consists of chancel, nave with north chapel and south porch, and west tower. The only remaining parts of the fourteenth-century church are the chancel arch, two windows reset in the north wall of the nave, the nave buttresses and a short stretch of nave wall adjoining the north of the tower. The tower was built in the late fifteenth or early sixteenth century. The church contains a twelfth-century font and sixteenth-century benches, and perhaps appropriately, in view of what is happening on the farm, a modern stained-glass window — the Benedicite window — depicting the seasons of the year and the creatures of land, water and air. Designed by A. E. Buss, it is a challenge to youngsters bored with the sermon to see how many birds, animals and fishes they can count!

The school, a Church of England aided school with a roll of about seventy pupils, serves not only Rushall but also five neighbouring villages and hamlets. Rushall was fortunate in possessing the best

The Benedicite window in Rushall Church

school building in the locality when many small village schools were closed a few years ago, so Rushall School, which has a very good local reputation, has become the centre for primary education in the area.

Rushall Farm actually comprises two farms, Rushall and Charlton, with the addition of a few fields from another. Rushall itself covers about 920 acres and Charlton about 650. In addition, some 80 acres on the southern boundary were bought in 1946 to round off the farm, making a total of some 1650 acres. (See map on pp. 4—5.)

The lower ground straddles the western branch of the River Avon (the one that rises near Marlborough, passes along the eastern side of Salisbury Plain, flows through Salisbury and reaches the sea at Christchurch), and stretches up to the edge of the Larkhill artillery ranges at about 700 feet above sea level. The soils vary from the heavy Blewbury in the vale to the light chalky Upton type with a fair share of flints on the hills. It is a steep farm with no fields being

A section of the north wall of The Manor, Upavon, showing the flint, chalk and brick construction

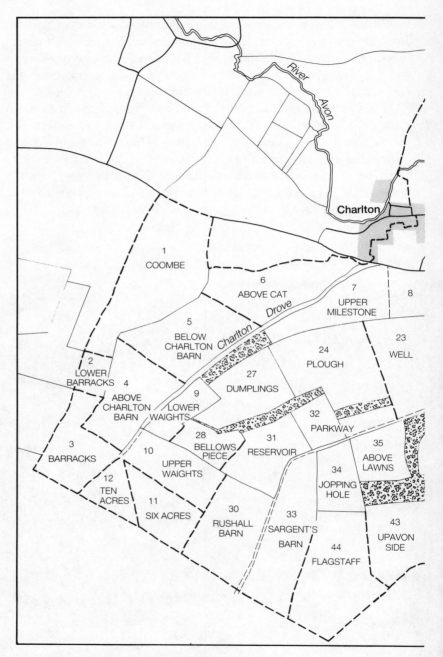

Map of Rushall Farm (Crown copyright reserved)

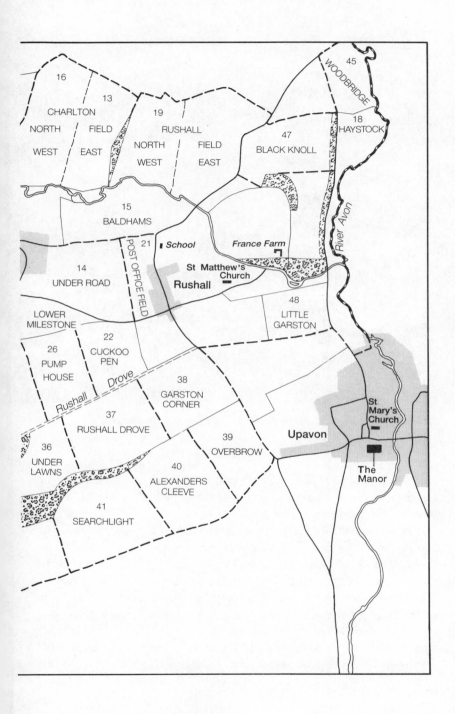

Sheep grazing a long ley; note the 'grass hedge' used for dividing
up large areas

entirely level and it is predominantly east- and north-facing, very
different from our rented farm on the other side of the valley, which
faces south and west.

The responsibility for Rushall passed into the Wookey family in
1945 when my father, Percy Wookey, bought it from Joe Maggs, the
co-founder of the old United Dairies (now part of Unigate). Prior to
that it had been in the Stratton family, a well-known name in
Wiltshire farming circles. The way it came into our family is
interesting. At the end of the war Joe Maggs suffered a disastrous
fire in the main set of farm buildings in the middle of the village. All
that was left of the cowsheds and barns was a brick-built dairy (still
standing) and Joe Maggs himself was very ill. After the fire he

summoned my father, who was then living and farming in Upavon, and told him that he was going to sell the farm and that he (my father) had to buy it! My father said he couldn't afford it — Joe Maggs said he would have to find the money — so he went away and consulted the bank. The upshot was that he did buy it, and Joe Maggs, who everyone thought was very near death, moved to another house in the village and without the worry of the farm recovered and lived very happily for another ten years or so!

From 1945 until his death in 1964, my father lived and farmed at Rushall. Because of the war years which followed the Depression of the 1930s, the whole place was in a pretty run-down condition when my father took it over. He started modernizing — at first the houses, none of which had the basic modern conveniences when he bought the farm, then the buildings and roads, and he made a start on the woods. Since 1964, this work has been continued and the farm is now reasonably well equipped. The main farm road is tarmac, the mill has been modernized and the cattle sheds and yards are all fairly new. Only the endless maintenance is left, and that seems to cost more and more every year.

In the days of Joe Maggs the farm was really an estate. It employed about sixty men, including carters, daymen, cowmen, shepherds, keepers, carpenters, wheelwrights and blacksmiths. Some of the old machinery is still in existence but no longer in use. There were, I think, two dairies, two flocks of sheep and sixty cart-horses with all their tackle. I am told that the last remaining thatched barn on the farm, the one at France (Francis Farm) that we have restored, used to be the foaling box for the cart-horses. It now serves as a bar when we have our sales of riding horses biennially in June.

Things are rather different today. Although we seem to have about seventeen names on the payroll, only about eight help the manager on the farm itself. In addition, we have a keeper-*cum*-estate-policeman, an estate maintenance man, a pensioner to keep the grass cut, a girl groom and a school leaver with the horses, the miller's wife part-time to help keep the mill records straight, a gardener and a part-time secretary. I am frequently told that we employ too many people and economically speaking I would agree. But there is more to life than economics and I feel very strongly that, provided you can afford it, those of us fortunate enough to be custodians of an estate such as ours, however small it might be, should accept the responsibility that goes with it and support as many families as possible, while accepting, of course, that numbers employed can never be as many as in the days before the war.

I am frequently asked whether the labour requirements on an organic farm are greater than those on a conventional farm. I think this question stems from the fact that people associate organic

Essential hand tools for the organic farmer: left, a thistle spud; right, a long-handled hook

An old steam cultivator, now pensioned off and standing
outside Rushall Mill

farming with old-fashioned farming and seem to think that we do
everything by hand and with horses. The impression may have
something to do, too, with the fact that we have between forty and
fifty horses on the farm at any one time, but this is a separate
venture, aimed at producing well-mannered young riding horses and
has nothing to do with motive power on the land! We do still run all
the machinery we had when we started converting — tractors, with
all their associated implements, combines, balers and, that most
useful but expensive tool, the rough-terrain fork-lift truck. We have
found that we require no more men to run the farm now than we did
before. We have no need for a man to spend weeks top-dressing (see
Glossary) corn and grass, no need for a man to operate a sprayer for
a month in the autumn and another two months during spring and
summer, and in the end we have less of everything we produce to

9

handle. We set out to establish whether an organic system of farming on our type of land was viable, and, as wages constitute a very high proportion of the costs of any system, it is doubtful if we could stay solvent if our system required us to employ twice the normal labour force, as many people seem to believe. Machines do the work of many men and, provided they are used with discretion and due care for the land, there is no reason to restrict or ban their use just because you eschew chemicals. A potato tastes the same whether it has been hoed by hand, horse or tractor!

Our livestock is made up of beef cattle, sheep and horses. The cattle are brought onto the farm either as ten-day-old calves which we then wean or as twelve-week-old weaners. In either event, we keep them for about two to two-and-a-half years before selling them as strong stores in the local market. The sheep are run as a joint enterprise and consist of a breeding flock of some 750 ewes (see chapter 7). The horses, like most horses, are not exactly fortune-makers, but a book I read early on in this venture said that on an organic farm at least three varieties of grazing animals were needed. This was sufficient excuse! We buy ten foals each year at weaning (September—October) and sell them at our biennial sales in June. We aim to produce well-grown, well-mannered young horses that people can buy with confidence to use in whatever discipline they choose.

The main enterprise is centred on the flour mill. We grow hard varieties of wheat suitable for bread-making, stonegrind it in the mill and sell it as organically grown wholemeal flour. The financial aspects of this and other enterprises will be discussed in later chapters, but at this point I would like to draw attention to two aspects that are a little unconventional. The first is the varieties of wheat that we grow. We started with the long-strawed variety Maris Wigeon, which although now officially outdated continues to be our number-one choice. It has never let us down, always mills and bakes well, and we have had many fields yield over 2 tonnes per acre. In addition, it makes ideal thatching straw when cut with a binder and combed or reeded by a special machine which delivers bundles of straight straw ready for the thatcher to use. There is a ready sale for any surplus bundles, as the local thatchers consider our straw, grown without chemicals of any sort, to be far superior to the conventionally

Young horses at grass

grown reed, and a 50 per cent premium is obtainable. As new varieties are introduced we fall for the publicity and try them, but to date have found nothing to last like Wigeon. Some may produce more for a year or so, but when the going gets rough, as in 1985, they tend to let us down by growing out in the ear (see Glossary), by shedding or by producing a poor sample. The other aspect concerns the question of testing bread-making wheats. If you sell to merchants, they subject it to an ever-increasing series of tests to establish whether it is suitable for milling. They test it for protein, for machinability, for its hardness and for its 'Hagberg falling number' (see Glossary). All these tests are designed to ascertain whether it is fit for milling into flour for bread-making. We have none of the

11

complicated machines necessary for performing these tests. Instead, we simply bake the flour, which, after all, is the test that all these other tests are about, and if it makes a good loaf it passes! Very simple and effective. What worries me is that, when we have had some of our samples tested officially they have often been declared 'unsuitable' for milling, yet in our bakehouse they produce perfect loaves. Are we in danger of becoming too sophisticated?

So the farm is a mixed one. In addition to the grass and wheat we also grow some kale and clean swedes for the sheep and cattle. Swedes may now be considered old-fashioned, having been replaced by silage produced with large quantities of bag nitrogen. But there is no better feed for young calves than chopped swedes and rolled oats, and whole swedes and straw make a very good ration for the older outliers. In short, we aim to be as self-contained as possible, with

The Manor, Upavon, our home since 1951

New calf shed at Charlton, built with Waney edge (see Glossary)
and old bricks

only the protein having to be brought in. All the dung made by the yarded cattle is returned to the land, and the system is, we hope, sustainable. In my view, no farm is worthy of the name if it is stockless — it merely exploits the finite resources of this planet, and that cannot be sensible.

As the farm is owner-occupied, the responsibility for the maintenance of all the buildings is a very real one. Over the years we have demolished quite a few old cottages, barns and buildings, but we are still left with a farmhouse, twenty-six cottages, one old thatched barn and some six sets of buildings. All these have to be maintained in good order and for this we employ an estate maintenance man, who also has to be capable of dealing with the estate water supply and drainage. We have adopted a livery of magnolia and blue for our thatched cottages — magnolia on the walls and blue paint on the windows and doors — with green and white for the field gates and the new buildings.

2

Why we converted

Before embarking on the technical aspects of converting a farm to an organic system I would like to try and explain why we have adopted this system at Rushall. It is a question I am always asked wherever I go (or by whoever comes to look round), so I feel justified in answering in some detail. This detail has to be something of a potted biography — not to publicize any virtues I may have, but to show how the decision came about, or, rather, evolved, for it took some twenty years to make.

I was born in 1925 in a village called Overton, near Marlborough in Wiltshire. In 1928 my father and grandfather took the tenancy of what is now Upavon Farms — some 3500 acres. These were, in fact, four farms — Cleeve, Upavon, Chisenbury and Littlecott — and were put out to tender by the owners, the War Department, who had bought about 100,000 acres of Salisbury Plain as an army training area at the turn of the century. All that land not immediately required for training was let as farms, and the four farms round Upavon were offered either separately or as one holding. It seems hardly credible now, but my father and grandfather were the only ones to tender for the whole block. There were a few offers for individual farms, but as my father and grandfather were the only ones to put in a bid for the whole they were successful. It was a brave decision in those days with the Depression about to hit the economy and farming facing very difficult times. I, and the rest of my family, now enjoy the result of my forebears' courage and foresight.

After preparatory school in Bournemouth, where I met the

headmaster's daughter, who later became my wife, I went to Malvern College from 1939 to 1945. There I took the first step towards being a doctor — a very lowly step, I might add, as we seemed to spend all our time cutting up frogs, cockroaches, smelly dogfish and even smellier rabbits sent to the school by my father in a biscuit tin! Then followed a very interesting and pretty ineffective career in the Army. I chose to be a sapper (a Royal Engineer to the uninitiated), as the corps ran a short (six-month) course at Clare College, Cambridge, which seemed to be a good idea at the time. Seriously, though, a sapper's training is really a wonderful start for any would-be farmer, as it is not all about demolitions and mine-sweeping. Learning about water supply, bridges, road-making and earth-moving equipment are all very useful on the farm, and you certainly learn about them as a sapper. When the Cambridge interlude ended we had to start soldiering in earnest, and Adolf Hitler obviously got wind of this as he shot himself just after we were commissioned. So we were shipped out to India to face the Japanese, who also saw us coming and decided to call it a day. With a high demobilization number in India there was no way of wangling early release, so we had three very happy years in India with servants to attend to our every need (my wife has never forgiven them!). The only blot on this colonial idyll was the rising tension between Hindus and Muslims in the run-up to independence — tension that spilled over into appalling carnage and produced the partition of the old Raj into what are now India and Pakistan.

We had left England for India in 1945 as wet-behind-the-ears second lieutenants, travelling by way of the newly re-opened Suez Canal in first-class cabins aboard the *Winchester Castle*. My one memory of that trip is having to lecture to battle-hardened troops from Europe — embarrassing to say the least! We returned as captains aboard the *Otranto* — stuffed a hundred to a mess deck in three-tier bunks — but who cared?

So back to England in 1948 to attend medical school for seven years — or to join my father on the farm. To his great relief (which I now understand) I opted for farming and joined him on the tenancy of Upavon Farms in 1949.

So began my farming career. The erstwhile captain RIE (Royal Indian Engineers) with servants galore now became a general dogsbody knowing a good deal less about farming than most of the men on the farm, and certainly far less than the foreman, Bob Dear. He had worked all his life on the farm, from cowman to tractor-driver to foreman, and I can honestly say that I learnt more from Bob than from anyone else, except, of course, my father. There is a world of difference between commanding a company of India troops and giving orders to a highly knowledgeable farm-worker. On the one hand you have King's (now Queen's) Regulations to bolster you; on the other, nothing but your own ignorance. But together we survived!

In 1950 I married the headmaster's daughter (who had waited for whom, we never have resolved) and in due course produced a family of one girl and two boys. Caroline, married to an American, lives in Delaware — ironically the home of the Dupont empire which manufactures many of the agrochemicals which we find so questionable; Nigel now runs the Upavon Farms and has three children to carry on (we hope) the family tradition; and Stephen, the youngest, is in the Church, now serving in London at All Souls, Langham Place, after three years in Paris, which we now have a good excuse to visit. Stephen is a cricketer and has two claims to fame — he opened the bowling for both Cambridge (1975 and 1976), and Oxford (1978), where he attended theological college after gaining his degree at Cambridge. The other is that he caught and bowled Geoffrey Boycott for a duck (no score) when Oxford were playing Yorkshire.

My entry into farming coincided with the introduction of the new agro-chemicals. Agroxone was the first we used — then a white powder that we scattered over the fields with a fertilizer distributor to control the charlock that always seemed to be present in corn on the chalk. It was marvellous. The charlock started curling up and very soon died and we really believed that we had found the answer. Then it was produced in 40-gallon drums as a dark-coloured liquid and we bought sprayers to apply it. This was even better — more acres per day and a more effective kill. And all this at the time we were receiving a subsidy cheque by every post as feather-bedded farmers! No wonder we really thought Utopia had arrived.

Throughout the 1950s more and more chemical sprays were introduced, and in common with most other farmers we used them

Filling a grass seed box with a long ley mixture

— at first to control the easy-to-kill annuals and then, as the sprays became more lethal, to control the docks, nettles and thistles. There seemed to be at least half-a-dozen new preparations every year and we all used them as fast as we could buy them. Farming had the weed problem licked!

During the 1960s three apparently unrelated cases of man's interference with nature came to my attention and set me thinking. The first concerned our wild English partridge. On the downs around Upavon we had built up a wonderful stock of these lovely birds. Starting in the shooting season of 1953 with a best day's 'bag' of 96½ brace (i.e. 193 birds) we had built up to a record day of 520½ brace in 1960. By 1970 our best day had reduced to 50 brace, and we have never had a good stock since. The build-up in stock coincided with the increase in the use of sprays, and in our ignorance − and, I must admit, arrogance − we said that the sprays didn't harm the partridges. We even went so far as to demonstrate that you could pour spray over a coop full of young partridges without harming them in the least. With hindsight I suppose the early sprays were nowhere near as efficient as the ones developed in the 1960s, which meant that they didn't destroy all the weeds in the cornfields as was the case later. As is now well known, a partridge chick depends on insects for the first three weeks of its life, so, while we were not killing the partridges directly, by destroying the weeds on which the insects lived we were effectively depriving the chicks of their vital food supply. A simple but deadly way of destroying partridges − not directly, but at second remove.

The second case was brought to my notice by Rachel Carson's book *Silent Spring*, which traced the death of vast numbers of robins in America to the use of DDT on elm trees. What happened was this. Dutch elm disease became established in parts of America and the Americans, not given to doing things by halves, sprayed every elm tree in sight with DDT. They knew that the disease was carried by the elm-bark beetle, so they thought that if they killed all the beetles the disease wouldn't spread. Simple logic! But nature is not that simple. In due course the leaves fell from the trees and were eaten by the worms. The amount of DDT on the leaves eaten by any one worm was insufficient to kill it, but it concentrated the poison in its body (bio-magnification) so that when the luckless robins ate the worms they soon acquired a fatal dose and died. Again, a case of death at second remove.

And the third case was the Thalidomide tragedy. It is still too clearly remembered for it to be necessary to go into details — suffice it to say that here was a drug which even after undergoing the usual screening for human medicines still resulted in such devastating effects. When the story was made known, largely through the *Sunday Times* 'Insight' team's book *Suffer the Children* (Deutsch, 1979), it became apparent that adverse data had been withheld by the manufacturer — data which, if made known to the medical committee responsible for passing the drug for human use, would have eliminated it immediately. So I thought: if a drug can be passed fit for people when it is potentially so dangerous, what may be the effects of the dozens (and there were dozens by then) of new agro-chemicals being used on farms so freely? Can we be sure they are safe? Have the long-term effects of these chemicals been studied? And where is the independent Ministry testing station?

Perhaps we should look for a moment at the introduction of a new agro-chemical. I have been told that it takes some seven years and £20 million to put a new agro-chemical on the market. The company who produces it does its own testing and before it can be sold it has to pass the scrutiny of the Pesticides Safety Precautions Scheme (PSPS). To quote the Ministry of Agriculture booklet *Approved Products for Farmers and Growers 1982*,

> The objective of this Scheme is to safeguard human beings, livestock, domestic animals, beneficial insects, wildlife and the environment against risks that could arise from the use of pesticide products. It is a formally negotiated agreement between the British Agrochemicals Association, the British Pest Control Association and the British Wood Preserving Association representing notifiers; and the Government Departments and Agencies responsible for Agriculture and Health and Safety in the United Kingdom.

Of the Agricultural Chemicals Approval Scheme, it says,

> This is a voluntary scheme under which efficiency of proprietary brands of agricultural chemicals can be officially approved. Approval cannot be given to a product containing a

new chemical or to a new use of an existing product until it has first been considered and cleared under the Pesticides Safety Precautions Scheme.

From this it will be seen that the PSPS is the more important body, controlling the introduction of all new agro-chemicals, and this leads on to the question of independent testing. Nowhere in the United Kingdom is there any independent Ministry testing station for agro-chemicals, so doubts as to the adequate testing of new products must persist.

Once you start thinking along these lines — the way chemicals affect ecological chains, setting up potentially destructive chain reactions, and the safety procedures regulating the introduction of new chemicals — you cannot help becoming more and more worried at the practices of modern agriculture. Remember, these thoughts were going through my head in the 1960s — now many more people are realizing that we may well be headed for catastrophe — and it was at that time that I decided that I would see whether it was possible to farm organically. To be fair, we at Rushall were in the fortunate position of having a lot of land, due, as I have said, to the foresight of my father and grandfather, and so we could afford to experiment. This work, I maintain, should have been undertaken by the Ministry of Agriculture, either by devoting one of its experimental husbandry farms to research into organic methods, or by conducting the same research on parts of all its farms. In spite of repeated efforts on my part (see chapter 11), and latterly by the organic movement generally, the Ministry has never faced up to the challenge. One day it will regret this.

There was a fourth factor that influenced me during the 1960s, and that was the recurring warning that the world's oil supply would give out in the near future. Over the last two or three decades we have been warned that oil production would cease to be economic by (take your pick!) the 1990s, the turn of the century, 2025 and virtually any other time before the year 3000! Experts seem to delight in trying to frighten the living daylights out of the general public, whether it be by warnings about meat, milk, fat, sex or television, and the oil experts are no exception. But, while their predictions should be taken with a pinch of salt, the fact remains that modern farming practices in the Western world are based on running down the world's fossil fuel resources to produce huge surpluses of food that cannot be used. Surely the situation is absurd. What is needed is a sustainable system of agriculture that is tailored to the needs of the twenty-first century, and that is what we are trying to establish at Rushall.

3

The farming year

What happens on a farm during the various seasons of the year? I find that many of our visitors, unless they are closely involved with farming themselves, find it difficult to relate the various activities with months of the year, and in addition there are many jobs that have to be done that are often taken for granted by the casual observer. I am often surprised by even local (non-farming) people asking me in June, for example, if we have finished planting the corn, or whether we have finished harvest! So, for the benefit of the reader without first-hand knowledge of farming, I offer here, before going on to describe in more detail what organic farming is and how we converted to organic methods, a month-by-month account of the farming year, and a framework within which our various activities may be set. (The Glossary offers further help with terms familiar enough to the farmer but not to the layman.)

The calendar starts with January, but traditionally the farming year begins at Michaelmas — 29 September — when most farm tenancies start. Indeed, the old custom of eating goose at Michaelmas is said to have originated in the tenant's practice of propitiating his landlord by giving him a goose when the tenancy came up for renewal. So I shall start with October and set out what happens on our farm at Rushall. Remember, we have no dairy and we use no chemicals, so the timing of various operations and certain specific events (such as calving) may well be different on other farms.

October

This is the month for drilling the winter corn and every man and tractor is kept very busy. We work the fields to prepare a seed-bed using ploughs, disc harrows, drags and seed harrows (see Glossary), and then wait for the weeds to start growing before planting the wheat or oats. 'Corn' is the generic name in England for wheat, oats, barley and rye, which are called 'small grains' in America, where the name 'corn' is reserved for what we call maize in England. And, just to add to the complications, corn planted in the autumn is called winter corn, and corn planted in the spring is called spring corn. Winter corn has to withstand the winter, and if it does, if the weather is not too severe, then it usually gives a better yield than that sown in the spring. You have to choose a variety that has been bred to withstand the winter if planting in October, as spring varieties would succumb in all but the mildest of winters.

On the stock front, October sees the first of the calves coming onto the farm. We buy ten-day-old calves from a calf group (see Glossary) in batches of ten to fifteen and wean them on skim milk powder. As we like to buy about 300 calves each year, we need to start in good time if we are to achieve our objective of having them all on the farm by Christmas. It is generally accepted that autumn-born calves do rather better than those born after the turn of the year, when, apart from anything else, the weather is usually a lot colder.

October also sees us collecting our batch of ten colt foals. This entails a lot of travelling, as I like to see the foal at home with its mother before deciding whether to buy. I usually place an advertisement in *Horse and Hound* and sit back and wait for the telephone to ring. We may also hear from any of our previous breeders who have a colt foal to sell, so we try to group the foals geographically and fix a day to see all those we want to see in a particular direction. It is surprising how long it takes, by the time you have found the place, seen the foal — which is often some distance from the house — and then done the deal (or not, as the case may be). When the foals have all been bought, they are grouped into two batches according to size, fed well and left to grow.

Spreading dung on a recently broken-up ley in September

November

We would like to finish drilling the winter corn in October, but inevitably the weather decrees otherwise and so the rest is done in November. The days get shorter and the land gets wetter, so less can be done each day than in October. However, the drilling is usually completed by the end of the month.

25

The rams are put out with the ewes so that lambing can start in April. As a rough guide, tupping — putting the rams to the ewes — on Guy Fawkes Day (5 November) means lambing on All Fools Day (1 April). The ewes are 'flushed' just before tupping, which means their diet is increased up to that date and then allowed to fall back. This is the time-honoured way of ensuring a good lamb crop, and, like all things in nature, it works sometimes!

The yearling cattle (i.e. the previous year's calves) are brought into their winter quarters and put onto their winter rations. This means hay and barley straw in the racks and rolled oats with a little home-made cake in the troughs. In addition, we use swedes straight from the field — chopped for the younger cattle, whose teeth are not strong enough to bite them, and whole for the older ones. The strongest cattle go to their winter quarters, which means a sheltered field where they can escape the worst of the weather, which can be very fierce on top of the Wiltshire downs.

Shooting and hunting start in earnest in November and we try to have the hounds about once a month. Foxes do little harm to game during the winter months; it is at nesting time that the keeper's patience is tried to the full, as game birds insist on nesting on the ground, where they are very vulnerable to all sorts of vermin — including the two-legged variety!

December

The winter routine of cattle-feeding is by now in full operation and this requires most of the men. They work in pairs and feed two or three lots each so that all the animals are fed in good time. Our cattle are kept in bunches of no more than fifty, so there are several lots to feed, and the swedes have to be lifted by a mechanical harvester whenever the weather permits. The Christmas and New Year holidays, with most of the staff having a week or so of their annual holiday as well, means that we seldom have a full staff to call on at this time of year.

Winter wheat competing with late snow (April)

January and February

I lump these two months together so as to dismiss them more quickly — quite the worst months of the year and the sooner they are over the better! Their one saving grace is that, as there is little to do on the land, we are able to have a small team working in the woods to keep them in good trim. I must say that a good bonfire in a wood that is being cleared is one of the better ways of spending a frosty February day! It is very satisfying to look back on what has been achieved during the day or during the week and to feel that that piece of wood will not need attention for the next five years or so. On the other hand, one of the worst jobs is thawing out the water troughs on top of the downs in a cold and freezing rain. This is something that the summer visitor seldom realizes has to be done, and it is certainly as different from walking the downs on a hot summer's day as can possibly be imagined.

Fencing is another job that tends to be done in the winter. We aim to fence each block of land bordering a road with permanent fencing, and sub-divide these blocks with electric fences. We only fence against cattle (the shepherd is responsible for his own fencing), and for the permanent ones we use cleft oak piles with three strands of barbed wire. In this way, if our cattle break through the electric wire, as they can do in a thunderstorm or driving rain, they should not be able to escape onto the roads. But we are all glad to see the back of January and February!

March

This is the month for the spring planting. Those fields not planted in the autumn with winter corn have to be sown as soon as the weather permits in the spring. This used to be in about mid-February, but for some reason or other the seasons seem to have slipped over the last few years and the land is seldom dry enough to be worked before March. The old saying that a peck of March dust is worth a king's ransom reflects this need for a dry March so that planting can be done in good time. Likewise, the saying that 'cuckoo oats' are no good also refers, indirectly, to the need for a dry March, because spring oats not planted before the cuckoo arrives seldom produce a decent crop.

Not only has the spring corn to be planted, but the winter corn all has to be rolled whenever conditions permit, and grass fields, used to produce hay, need to be rolled with a heavy roller to push the flints, and other stones, well into the surface so that the grass-cutter does not get damaged. Nothing is more frustrating than to have the mower out of action at the busy time of hay-making because of an argument with a flint!

March is a difficult month for the stock. They are all scenting the regrowth of grass — those inside the yards can't do much about it, but any outside keep pushing for that little bit of greener grass the other side of the fence and, unless the fences are in good order, the cattle break them and escape. It is always difficult getting them back into their winter field — having tasted the fresh grass while roaming free, they are very loath to go back to the bare pasture of their winter quarters. All animals at this time of year tend to look their worst — lice seem to appear from nowhere on housed animals unless they are regularly dusted with powder, and any warble-fly eggs that were laid the previous summer now hatch into larvae that burrow under the skin on the backs of the cattle and have to be dealt with. Food for the animals is running short and this causes a lot of worry if the season is late. It is a good rule of thumb to have half your winter food for the animals still in the barn on 1 February each year, as all the animals are growing and eating more as the winter goes by. Easy to say and obvious, but we have all been caught at one time or another with empty barns and no grass!

Freshly rolled wheat in the spring

April

April is traditionally the month for lambing and so a very busy month for the shepherd, who has to contend with anything up to fifty or sixty ewes lambing each day (and night). If there are 1.75 lambs per ewe, which is a good average, this means he has to make sure that about one hundred lambs know which is their mother and are strong enough to feed from her. He must make sure too that the ewes lamb properly, that there are no complications, and that they have enough milk for their lambs. No wonder shepherds at this time of year look drawn and haggard and are prone to bouts of ill temper! Many people seeing lambs frisking in the spring sunshine do not realize just how much effort has to be put in by the shepherd to produce this pretty sight.

The grass should be growing during April, so this is usually the month for turning out the cattle. On an organic farm, spring growth depends on the micro-organisms in the soil, which must release the nitrogen needed for growth to start, and these organisms do not operate until the soil warms up. On a 'chemical' farm this process is replaced by bag nitrogen — a chemical fertilizer that releases nitrogen very rapidly so that the grass can start growing earlier. Thus, turn-out usually occurs some ten days to a fortnight later on an organic farm than on a conventional one, and the period is all in all a difficult one for the organic farmer!

On the arable side, any field work not done during March is now completed. April is also the month for undersowing the corn with a ley mixture (see Glossary) on those fields that are due to be grass next year. The grass seeds are broadcast onto the surface when the corn is about 4—6 inches high and then rolled in. They germinate and start growing, but the corn soon grows so tall that their development is checked. After harvest, the seedlings, sometimes so small that they can barely be seen, are exposed to full light and they quickly develop. If there is enough rain after harvest, they can be grazed by the sheep in the autumn, but, in any case, by the following spring there should be a good grass field.

Eight to nine month old Hereford × Friesian steers on summer grass

May

This is the month of the 'blackthorn winter' (see Glossary) and a month for organic farmers to go away fishing! The growth of all crops tends to be slow, and seeing one's neighbours' fields, with many units of bag nitrogen to help them, growing away so much faster that one's own is enough to try the patience of Job. The stock seem to be chasing every blade of grass as soon as it appears, the hay fields can scarcely conceal a hare, and the corn seems to grow downwards rather than up. Yet the cattle, sheep and horses seem content with their lot, the corn *does* get a little taller, and the

31

countryside, as soon as the blackthorn winter is over, looks at its best. Could it be that our grass, unforced by artificial fertilizers, contains more nourishment than the lush-looking stuff on our neighbours' fields?

Mangolds are planted in May with a precision drill that is supposed to plant one seed every 6 or 9 inches. The machine does its job well, but unfortunately the mangold has not been told about it, as in each mangold 'seed' there are one, two or even three viable seeds. Thus, while the machine plants one seed at the required spacing, that seed often throws up two or three seedlings, which means the crop has to be hand-hoed, an occupation that rates very low on most farm-workers' list of preferences! For this reason there are very few mangolds grown nowadays, but we like to grow a few acres as they are very useful for the cattle in February, March and April before turn-out. Yarded calves will leave their cake for mangolds and they sound like a room full of kids eating apples. It must be a nice change for them to have something juicy after all the dry hay and straw they have been eating for so long.

As soon as the cattle are out of their sheds and yards, dung-carting starts. What used to be a backaching, seemingly endless, task when done by hand is now all done mechanically with a mechnical loader and tipping trailers. We put all the dung in heaps on or near fields that will need it in the autumn and it quickly rots down during the summer. If we have time, we turn the heap once to allow more air to get in to help break it down further, as well-rotted dung is what we require for our fields. Even if we don't have time to turn it, the one move from yard to heap effects a fair breakdown so that when it is spread in the autumn it covers the ground well and is beneficial.

June

June was once the month for hay-making. It still is on some farms, but we find that, although we make some hay in June, most of it is made in July. Wimbledon fortnight is usually our best time and we often go on making hay until the corn harvest is ready. Early, young grass makes the best-quality hay, if you can get the sun to dry it. It takes more drying than anything cut after the longest day, when the

Cutting grass for hay with a modern disc mower

sap starts going out of the grass, making it easier to dry but not improving the quality. We tend to make some early hay for the baby calves, who need the high protein contained in young grass, and the main bulk after the longest day to provide the quantity needed for the older cattle and horses.

Shearing the sheep takes place in June — ideally, when the new wool has started to grow and the grease makes shearing easier. The shepherd used to do it all himself — he can shear 120 sheep a day on his own — but recently he has taken to having a team of shearers come in to do the job more quickly. Years ago we would cut stinging nettles to put on the straw in the barns where we used to house the sheep overnight to ensure they were dry in the morning

33

(you can't shear a wet sheep). This prevented the straw from becoming entangled in the wool, which meant a lower price, and I can still remember the distinctive smell of the nettles after the sheep had been there all night. Now the shearers are more patient and wait for a fine day!

June is also the month for sowing the swedes and kale we grow for the cattle and sheep. These crops are very susceptible to the 'fly' — the flea beetle — which devours the seedlings as they appear and can decimate a crop. The modern method of control is to dress the seed with an insecticidal powder, which somehow copes with the problem very well, but on our system we have to rely on an old-fashioned dodge that the old men on the farm told me about. The fly is most destructive during the hot, bright days of early June, so the trick is to delay planting the swedes and kale so that the seedlings do not appear before the longest day. This means, in effect, planting about

Turning hay with a side-rake

the third week of June, as these seeds germinate very rapidly given a good shower after sowing. In this way, the destruction caused by the fly is avoided and, with swedes, the likelihood of a mildew attack later on is very much reduced.

July

This month sees the end of hay-making and quite often the beginning of harvest. The first crop for the combine is winter barley, which is usually ready by about mid July, although this, of course, varies from season to season depending on the weather. The normal sequence of combining runs as follows: winter barley, spring barley, winter oats, winter wheat, spring oats and lastly spring wheat, which will probably be ready in early September. Modern varieties of corn tend to be ready to harvest about a month earlier than in earlier times, when September, with its harvest moon, was the harvest month.

July is our busiest month for receiving visitors. There is little to see on any farm in the winter, so we tend to concentrate the visits into June and July, leaving August free for harvest. We have had many visitors, the most distinguished being HRH Prince Charles in 1984, and they include a growing number of genuine farmers. There are also the naturalists, the conservationists and the young farmers, and to save a lot of talking we produce a hand-out (see Appendix A), giving the facts about the farm. The two things that most of our visitors remark on are the cleanliness of the crops and the abundance of wild flowers. We are particularly proud of our orchids, whose seed is so minute that it is airborne — it does not need a wind — and I am told it takes seven years to develop into a flowering plant. A list of wild flowers found on the farm in a survey carried out by a student in 1983 appears at Appendix B.

August

August is the month of harvest and all that goes with it — combining, baling, mechanical breakdowns, drying the corn and ploughing for

Baling hay

next year's harvest. This very busy time marks the culmination of the arable farming year, when all the work of the preceding months achieves fulfilment. Ideally, corn should be harvested when its moisture content is 14–15 per cent, at which level it does not require drying. Unfortunately, not many English summers allow you to harvest all your corn at such a low moisture, so it has to be dried artificially — a costly process but vital, as corn with a moisture content of more than 15 per cent will not keep. It tends to heat when put in the silos, and once it starts heating it very soon is spoiled and becomes unsaleable. We try not to harvest anything at a moisture level over 20 per cent, but in a season such as 1985 one is thankful to be able to harvest at all and moisture levels up to 25 per cent are not uncommon. This has to be brought down very quickly — within a day — to a reasonable level (about 17—18 per cent), and then in a second drying process reduced to a level safe for storage. This can be very costly and frustrating, as when you have the combines ready to go into something much drier you often find yourself having to wait while you redry some corn that was cut very wet. Trying to dry from a high moisture level with high temperatures is risky — not only does it destroy the germination, but also it seldom achieves a uniform moisture, with the result that the heap of corn quickly starts spoiling.

All the straw produced on the farm is recycled — either by way of food for the cattle or as bedding. Barley straw, especially spring barley straw, is a very useful fill-belly for cattle, as is oat straw,

Combining wheat

although this has the disadvantage that pieces of oat chaff may lodge in the eye during windy weather. The chaff, the outer covering of the grain when growing, is easily blown around and is of a shape to fit neatly on the eyeball. 'Oat hud in the eye' is what we call the condition on the farm and if it is not removed very quickly it creates an abscess and eventually causes blindness. The easiest way to remove an oat hud is with a bit of dirty grease on your finger tip — the oat hud will stick to it and out it comes. Easier said than done on the windy Wiltshire downs in winter! The wheat straw is used for bedding — it is not palatable — and makes very good farmyard manure.

A full barn ready for the winter

September

Quite the best month of the year! Harvest is over; the weather is nearly always perfect, with misty dawns and hot days followed by refreshing evening breezes. Maybe I am biased, as I was born in September! On the farm, the barns are all full, the lambs are ready for sale, and we are getting ready for another year, another harvest. The dung is taken from the heaps and spread, the stubbles are ploughed and made ready for drilling and the harvesting machinery is all cleaned and stored away for next year. It is the time of year to look back on the recent harvest and resolve to do better for the next one. There is a slight feeling of anti-climax when the last field has been cut, but this quickly gives way to a feeling of anticipation for the year ahead. I like September!

And so we start again.

4

What is organic farming?

Some of the main features of organic farming have already been alluded to in earlier chapters. Here I shall attempt to provide a fuller picture of what organic farming is and what it entails. The very word 'organic' is something of a problem, conjuring up, for some people, a vision of long-haired, sandalled enthusiasts not entirely in their right minds. 'Sustainable' is perhaps a better description of the type of farming concerned, as it conveys what we are trying to achieve without the emotive overtones.

'Going organic' is really a philosophy of life as much as a system of farming. Because farming organically is both more difficult and less remunerative (at least initially) than conventional farming, no one should embark on it without fully realizing and being prepared to accept that it involves a lifetime's commitment — you cannot be organic one year and conventional the next. It takes at least ten years (we took fifteen) to convert a farm from one system to the other, and during that time there will be many problems to overcome, so, unless the will and determination are there in the first place, the temptation to slip back into conventional methods will probably be too great.

Organic farming means different things to different people. Sir Derek Barber, Chairman of the Countryside Commission, who one would expect to have a feel for it, said in an article in *Power Farming* (November 1982), 'I find the arguments on behalf of organic farming wholly unconvincing . . .' and went on to use such emotive phrases as 'muck and mystery', 'a scent of humbug', 'delights of dung' and

A 'grass hedge' with the chicory in flower

'organic antics'. This fairly accurately reflected the attitude of the agricultural and conservation establishments, but things have changed a great deal in the years during which we have been trying to work out a sustainable system, and I think there is now a much more sympathetic understanding of what organic farming is trying to achieve.

A simplistic definition of organic farming might be 'farming without chemicals'. Indeed, during the conversion of our own farm we used the term 'chemical-free' for all fields converted to the system. But this does not really cover the concept of the sustainability of organic farming, nor does it convey anything of the philosophy behind it, which, as with all things in nature, becomes more complex

42

the more you think about it. In fact, the philosophy behind organic farming concerns nothing less than the complete cycle soil—plant—animal—man.

There have been many attempts to define organic farming. One of the more succinct definitions was offered by the United States Department of Agriculture in 1981:

Organic farming is a production system which avoids or largely excludes the use of synthetically compounded fertilizers, pesticides, growth regulators and livestock feed additives. To the maximum extent feasible, organic farming systems rely on crop rotations, crop residues, animal manures, legumes, green manures, off-farm organic wastes, mechanical cultivation, mineral bearing rocks, and aspects of biological pest control to

Turning a dung heap to achieve a better breakdown

maintain soil productivity and tilth, to supply plant nutrients and to control insects, weeds and other pests.

That is one of the shorter definitions! Or you may prefer the one published by IFOAM (the International Federation of Organic Agriculture Movements), according to which the objectives of organic farming are

1 To work as much as possible within a closed system, and to draw upon local resources.
2 To maintain the long-term fertility of soils.
3 To avoid all forms of pollution that may result from agricultural techniques.
4 To produce foodstuffs of high nutritional quality and sufficient quantity.
5 To reduce the use of fossil energy in agricultural practice to a minimum.
6 To give livestock conditions of life that conform to their physiological needs and to humanitarian principles.
7 To make it possible for agricultural producers to earn a living through their work and develop their potentialities as human beings.

The Soil Association, which was founded in 1946 by Lady Eve Balfour, and which embraces organic farming, says that its role is 'to promote a fuller understanding of the vital relationship between soil, plant, animal and man'. It 'believes that these are part of one whole, and that nutrition derived from a balanced living soil is the greatest single contribution to health (wholeness). For this reason it encourages an ecological approach and offers organic husbandry as a viable alternative to modern intensive methods.'

It is highly significant that there is no official definition of organic farming by the British Ministry of Agriculture, Fisheries and Food. In America, as we have seen, there is one, as there is in France, Switzerland, West Germany, Holland, Austria, Belgium and Israel. Could it be that our own farming establishment has become so closely involved with the agro-chemical companies, on whom they depend for advertising in trade papers, research and development, as well as free advice, that they have become blind to the potentialities of modern organic farming? When asked in July 1986 if it had a definition, the Agriculture Development and Advisory Service, a Ministry department, replied,

> We have not thought it helpful to formulate a separate definition from those already in use. Our aim in ADAS, in relation to our contacts with individual farmers, is to provide them with advice to help them to achieve their own objectives. We of course recognise that these objectives can embrace a wide range of factors in addition to economic performance.

This seems to endorse the view of a chief fisheries officer who said to me recently, 'The Ministry is only concerned with the production of food, not with nature'.

So, take your pick! Maybe 'farming without chemicals' has something to offer, if only simplicity. Or consider my wife's version: 'natural farming'.

In effect, organic farming means no chemical sprays and no soluble fertilizers. What organic farmers try to achieve is a healthy soil, as they believe this will produce a healthy plant that can withstand attacks by various diseases, so eliminating the necessity for sprays. To achieve a healthy soil it may be necessary to add insoluble substances such as ground North African rock phosphate, lime or basic slag (see Appendix C). These items, being insoluble in water, do not feed the plant directly as bag fertilizer is designed to do: rather, they balance the soil so that the soil organisms can go to work on them and break them down into a form that is available to plants.

The Soil Association's *Standards for Organic Agriculture* (June 1987) governs the conduct of organic farming in Britain by permitting the sale of produce under the Association's Symbol only when it has satisfied the conditions laid down in the Standard. As a comprehensive statement of what organic farming is all about it is unequalled and those parts of it relevant to our system at Rushall are, with the kind permission of the Soil Association, reprinted here and in Appendix C. The full document may be obtained from the Soil Association, 66 Colston Street, Bristol BS1 5BB, price £5.90 (inc. p & p).

The start of a bastard fallow after one pass with a chisel plough

Foreword

The Soil Association is a registered charity. Founded in 1946, it works both nationally and internationally to promote a better understanding of the links between organic agriculture, care of the environment, food quality and human health.

The Symbol Scheme

The Soil Association Symbol Scheme was established in 1973. Its principal aims and objectives are:

1 To define an organic farm management practice which produces high quality food, is sustainable, avoids damage to the environment, and is based on ethically sound principles.
2 To establish and promote a well recognized quality mark and

thereby offer the consumer a means of purchasing genuine organically grown produce.

3   To protect the consumer from fraudulent trading.
4   To protect the farmer from unfair competition.

It is now Britain's foremost organic quality standard with over 500 Symbol Holders.

The Soil Association Symbol is a generic quality mark. It is awarded, under annual licence, to:

1   Farmers and growers producing certified organically grown foodstuffs.
2   Processors and manufacturers of foodstuffs of organic origin.
3   Manufacturers of fertilizers and other inputs which conform to the production standards.

The Soil Association Standards define the production system. They lay down guidelines specifying rotational, management and welfare practices, as well as regulations and limitations for use of fertilizers, agro chemical sprays, animal feedstuffs and veterinary products.

1.   Introduction

Organic (biological) agricultural and horticultural systems are designed to produce food of optimum quality and quantity. The principles and methods employed result in practices which:

co-exist with, rather than dominate, natural systems;
sustain or build soil fertility;
minimize damage to the environment;
minimize the use of non-renewable resources.

The enhancement of biological cycles, involving micro-organisms, soil fauna, plants and animals, is the basis of organic agriculture. Sound rotations, the extensive and rational use of manure and vegetable wastes, the use of appropriate cultivation techniques, the avoidance of fertilizers in the form of soluble mineral salts, and the prohibition of agro-chemical pesticides form the basic characteristics of organic agriculture.

What is organic farming?

The Soil Association Standards apply to foodstuffs grown by producers who have followed the above principles and who have undertaken to adhere to the standards set out in this document.

These Standards are under constant review, and the Soil Association reserves the right to amend them from time to time. The Soil Association acknowledges the urgent need for further research and development work to evolve Standards for organic agriculture in the future.

Explanation of terms used in the text

Symbol Standard   Production that meets all the Livestock, Arable, Horticultural or Processing Standards. Prohibited chemical inputs may not have been applied during no less than the last 24 months for grassland and during a period of 24 months prior to planting for arable and vegetable crops.

Dung spreading

In Conversion   Conversion must be effected according to a progressive plan approved by the Soil Association that encompasses the entire holding or a physically and financially separate section of the holding. Production on land which has been part of an agreed conversion programme and which has had no chemical inputs for at least six months for grassland and during the six months prior to planting in the case of arable crops will qualify for conversion status.

All produce and crops not of Symbol Standard or in conversion will be regarded as conventional.

Recommended, Permitted, Restricted, Prohibited   Throughout this document, these headings are used to distinguish between different practices:

Recommended   Fully recommended as good management practice.
Permitted   Allowed to be used in Symbol Standard production, subject to any qualifications listed.
Restricted   Practices which are not fully compatible with organic principles, and therefore should not constitute a major part of the organic system.

These will: *Either* have qualifying conditions attached to their use, as detailed in the text, *or* can only be used with the specific permission of the Soil Association Symbol Committee, if no conditions are appended to them in the text.
Prohibited   Practices and substances not permitted for Symbol Standard production. Use of prohibited substances may result in withdrawal of the Symbol for two years.

Conversion from conventional production systems   The conversion of a field or holding from conventional to organic production must be carried out according to a plan agreed with the Soil Association. The plan must be designed to result in a viable and sustainable organic system operating to full Symbol standards. It must include proposals for: (a)   a rotation which balances fertility building and exploitative phases; (b)   appropriate manure management; (c)   appropriate cultivations.

Conversion must be undertaken on a farm or part of a farm large enough to allow a viable organic rotation. Prohibited inputs may not be used at any stage during the conversion.

The conversion programme must begin with the fertility-building phase if the land was previously under exploitative cropping. During conversion, livestock should meet full Standard requirements for welfare and veterinary treatment.

The conversion must be monitored by the Soil Association on at least an annual basis.

Produce may only be sold carrying the Soil Association Symbol after a conversion period of at least two years. Under certain conditions, and with reference to chemical residues, the Soil Association may extend or reduce this period.

Produce may only be sold as 'in conversion' after inspection of the farm, approval of the conversion plan and after the time limits that are stated in the conversion standards have elapsed.

2.   General Production Standards

Soil Management

Appropriate soil management is fundamental to successful organic production. The development and protection of optimum soil structure and fertility is the main goal of such management.

An optimum soil structure can be described as 'a water-stable, organic enriched, granular structure where all the water reserves within aggregates will be large enough to allow rapid drainage, to admit air and to facilitate the deep penetration of roots' (Elm Farm Research Centre: *The Soil,* 1984).

The development of such a structure relies partly upon natural physical and biological processes such as cracking, weathering and the activities of soil organisms, and partly upon management.

Management should ensure: (a)   regular input of organic residues; (b)   a level of microbial activity sufficient to initiate the decay of organic materials; (c)   conditions which ensure the continual activity of earthworms and other soil stabilizing agents; (d)   as far as possible, a protective covering of vegetation, e.g. green manure or growing crop; (e) appropriate cultivations.

Mechanical cultivations can initiate rapid improvement in soil structure, although the effect will be temporary unless it is reinforced by structuring through biological activity. Appropriate cultivations should achieve: (a) deep loosening of the soil; (b) minimal surface disruption; (c) timeliness to ensure appropriate tilth and to avoid damage to existing structure.

Rotations

Sound rotations achieve a balance between fertility-building and exploitative phases. They should aid the maintenance of soil fertility, soil organic matter levels and soil structure, whilst ensuring that sufficient nutrients, particularly nitrogen, are available and nutrient losses are minimized. Rotations are the primary means of minimizing weed, pest and disease problems. Whilst there cannot be a definitive rotation, the following guidelines should be observed:

(a) a balance should be achieved between fertility-building and exploitative cropping;
(b) deep-rooting crops should be alternated with shallow-rooting crops;
(c) alternate high and low root-biomass* crops;
(d) alternate nitrogen-fixing and nitrogen-demanding crops;
(e) weed-susceptible crops should be alternated with weed-suppressing crops;
(f) time intervals should be followed which are appropriate for similar disease-susceptible or disease-host crops;
(g) green manures, catch cropping and undersowing should be utilized to ensure maximum soil cover.

These guidelines are most easily followed within a mixed ley farming system.

Permanent grassland containing clover generally provides a stable system of nutrient supply and also weed, disease and pest control.

Special consideration, however, must be given to rotational aspects of systems that are predominantly arable or horticultural. The replacement of off-farm inputs by management of internal processes, which is sought in organic production, is more problematic on these holdings than on mixed farms.

Whilst these problems need to be accommodated, it is desirable that such holdings should move towards a better balance between fertility-building and exploitative management, and away from a total reliance upon outside inputs. Rotations which include legumes and maximum amount of green manuring and catch cropping are essential in these systems.

It is recognized, however, that the greater diversity of cropping and the encouragement of predators which is a feature of small intensive specialized horticultural holdings in part reduces the need for formal rotations.

In predominantly arable and horticultural systems the following management practices apply:

Recommended Rotation and manure management systems in accordance with the guidelines set out above.
Permitted Rotations falling short of these guidelines, but utilizing legumes, green manures and catch cropping. Protected crops — continuous cropping of the same genus — is allowed provided that pests and diseases are controlled by the methods outlined in this document. Perennial crops are allowed provided nutrient supply, weed, pest and disease control are effected by the methods outlined in this document.
Restricted Alliums, brassicas and potatoes, as outdoor crops, should not return to the same land until a gap of 48 months has elapsed from planting date to planting date. Green manures are excepted.
Prohibited Cropping systems not defined above which rely solely on outside inputs for nutrient supply and weed, pest and disease control.

For the Soil Association's List of Recommended, Permitted, Restricted and Prohibited Substances and Practices, see Appendix C.

3. Livestock Husbandry Standards

Introduction

On many organic farms, livestock enterprises using grass/legume swards form an essential part of the fertility-building phase of the rotation. Thus standards for organic livestock must be considered in the context of a whole farm or farming system which is being managed organically. Farmers applying for the Symbol for livestock enterprise must therefore also comply with the General Husbandry Standards.

Layout of the Livestock Husbandry Standards

The Soil Association Livestock Standards apply to housing, welfare, feeding and veterinary aspects of livestock husbandry. In Section 3, the general principles governing each aspect of organic livestock management are outlined.

Record Keeping

Both physical and financial records of animal enterprises may be requested for inspection. The inspectors may require to see records of: (a) bought-in stock listing age and source of purchase; (b) veterinary treatment (see 'Veterinary below'); (c) purchased feedstuffs (see 'Livestock nutrition' below).

Principles of Organic Livestock Management

A. Welfare and Housing

The general conduct of animal husbandry should be governed largely by physiological and ethical considerations, having regard to behavioural patterns and the basic needs of animals. The European Convention on Farm Animals requires that they should be kept according to their physiological and ethological needs. This would rule out cages, tethering and penning (except for short periods), unsuitable feeding, growth promoters or other interference with the normal growth pattern. The MAFF publication *Farm Animal Welfare* lays down codes of practice for farm animals covering both welfare and housing considerations. The Soil Association Standards will in some cases specify more stringent standards, but will adhere to the MAFF guidelines as a minimum. Outline principles are as follows:

1 The permanent housing of breeding stock is prohibited.
2 The prolonged confining or tethering of animals is prohibited.
3 Herd or flock size (either too large or too small) must not adversely affect the individual animal's behaviour patterns.
4 Stock must have access to fresh water at all times.
5 Buildings for housing livestock must have adequate natural ventilation and lighting, and allow sufficient room for the free movement of stock.
6 Materials used for housing must not be treated with paints or wood preservatives toxic to stock.
7 Bedding materials must be provided.
8 All stock should have access to pasture during the grazing season unless specifically excepted by the Symbol Committee.

B. Livestock Nutrition

Sound nutrition forms the basis of the health and vitality of farm livestock. Thus organically grown feedstuffs fed in correct proportions are the basis of Symbol standard requirements. Particular attention should be paid to the physiological adaptation of livestock to different types of feedstuffs.

Ration Formulation   For ruminants, forage should constitute no less than 60 per cent of the total daily dry matter intake. Ideally the farm itself should provide most of the Symbol standard forage and concentrates for its livestock. Because of practical difficulties being experienced by producers in obtaining 100 per cent Symbol standard rations, and in line with current IFOAM guidelines, certain variations have been made to these principles in the standards. These include allowances for feeding specific quantities of 'conventionally produced' and 'in conversion' feedstuffs.

They are permitted as follows:

A minimum of 50 per cent of the total dry matter intake *must* be full Symbol standard.

A maximum of 50 per cent of the total dry matter intake may be made up of 'in conversion' feedstuffs.

A maximum of 20 per cent dry matter intake may come from pre-conversion or conventional sources.

These requirements will be assessed on annual total dry matter intake, but attention will also be given to intake on a daily herd basis. The proportions apply to the total ration. However, grazing must only be on pasture that is of Symbol status or 'in conversion'.

Non-organic feeds will be disallowed totally when improved availability of organic feedstuffs renders this supplementation from conventional and 'in conversion' systems unnecessary.

The Symbol Committee may require bought-in conventional feeds to be tested for residue levels, at the producer's expense.

Mineral and vitamin supplementation On well established organic farms, sound nutritional practices should render mineral supplementation unnecessary. Where there is a known dietary deficiency in home-grown feeds, or as a result of soil deficiencies, supplementation will be permitted. Minerals and vitamins fed must be from approved sources only.

It is strongly recommended that all stock should have access to green/fresh fodder on a daily basis as available.

Permitted Additions to the diet of naturally occurring mineral/vitamin rich supplements (e.g. seaweed, bonemeal, cod liver oil, yeast).

Restricted Mineral licks — only those with flavour enhancers and other non-mineral additives, including urea. Straight mineral salts and synthetic vitamins in feed — only in cases of a known farm deficiency.

Prohibited   Routine addition to feed of restricted mineral and vitamin supplements. Mineral licks with flavour enhancers and other non-mineral additives. Urea.

Additives and medications   Naturopathic probiotics may be permitted as feed additives, but only with the specific approval of the Symbol Committee. All other food additives and in-feed medication are prohibited.

C. Veterinary

Central to the approach of organic livestock husbandry is the prevention of disease. Health in farm animals is not simply the absence of disease, but also the ability to resist infection, parasitic attack and metabolic disorders, as well as the ability to overcome injury by rapid healing.

Animal health results directly from a combination of good management practice, sound nutrition and good stockmanship. Thus veterinary treatment must be considered as an addition to, and not a substitute for, good management practice.

If illness does occur, the aim must be to correct the imbalance which created the disorder, rather than simply to deal with the symptoms of the illness alone. Complementary or alternative treatments should be used in preference to conventional treatments.

The Soil Association Standards will be updated in the light of new developments in alternative treatments.

Conventional treatment   Conventional drug treatment is permissible in the following circumstances: (a)   in order to save life; (b)   to prevent unnecessary suffering to an animal; (c)   to treat a condition where no other effective treatment is locally available.

Treatment of healthy animals is prohibited. This includes routine use of prophylactic drugs.

Growth promoters and hormones   All growth promoters and hormones for fertility disorders, heat synchronization, production stimulation and suppression of natural growth controls are prohibited. In the rare case of the need to induce parturition for medical reasons, then a natural prostaglandin may be used.

Worm fluke and husk treatments   Clean grazing systems, to prevent the build-up of unacceptable worm burdens, are recommended.

Specific treatments may be administered where stock are known to be carrying unacceptable worm burdens. In these circumstances strict identification procedures and withdrawal periods must be observed or the animal must be withheld from sale as Symbol standard.

The routine use of anthelmintics is restricted to cases where a known farm problem exists, and only then with the specific approval of the Symbol Committee.

Vaccines   Vaccines may only be used where a known disease exists on a farm which cannot be controlled by other management means. The approval of the Symbol Committee must be obtained in these circumstances.

Other drug treatments   All other conventional veterinary drugs are prohibited for routine use unless specifically excepted or if the approval for use of a drug is specifically obtained from the Symbol Committee.

A sample of loaves baked in Rushall Mill

Record keeping and withdrawal periods Conventional treatment must be subject to careful record keeping and observation of strict withdrawal periods.

The following records must be kept, clearly identifying the animals or groups of animals concerned.

(a) All sick and treated livestock.
(b) All conventional veterinary treatment, including details of the condition, the treatment used and its duration, as well as the brand name of the drug and the manufacturer.
(c) Any treated animals must be individually identifiable during the drug withdrawal period.

Withdrawal periods after treatment with conventional drugs are one month or three times the manufacturers recommendations, whichever is the longer.

D. Bought-in Stock

It will not always be possible to obtain purchased stock from organic sources, although farmers will be expected to show that they have made reasonable efforts to do this. Conventionally managed stock may be bought in at present. Bought-in fattening stock should receive organic treatment from the point of parturition if they are to be marketed as Symbol standard.

Recommended Closed breeding herds/flocks.
Permitted Worming of bought-in stock upon entry to the farm.
Prohibited Purchase of calves and fattening stock from livestock markets.

E. Conversion

Farmers will be expected to show that they have fufilled the requirements in the 'General Husbandry Guidelines' with regard to non-use of chemical and fertilizer inputs. They will also be expected to show an understanding of the approach to veterinary treatment outlined above. Evidence will be required of the adherence to the above veterinary standards in the last two years before the Symbol application.

5

How we converted

Converting *yourself* to the idea that organic farming is the only sensible way to farm for the future is the biggest hurdle to overcome when you start looking into the subject. You will soon find plenty of reasons for *not* changing — returns, markets, advice from others, length of time required, weeds, the uncertainties of taking a completely new approach, the lack of guarantees — but once you have finally made the decision it is surprising how the difficulties and problems become progressively less worrying.

So how do you go about it? I shall set out what we did — not as the correct or best way, for, as will be seen, we made our share of mistakes — but as a guide to anyone thinking of changing. As we proceed I shall indicate where I think we went wrong, and why, and how it might be done better.

The map of the farm in chapter 1 shows the position and numbers of the various fields, and the cropping list in table 5.1 gives the cropping for the harvests of 1986 and the two previous years, with the year each field was first 'chemical-free'. This table should be read in conjunction with the following account of our problems in conversion.

We started out in 1970 by scheduling two fields — field 48, called Little Garston, and field 35, Above Lawns. Little Garston was chosen as one of the best fields, and Above Lawns as *the* worst field, on the farm. The reasoning went like this (remember we were

Table 5.1 Cropping 1984—6 (with year each field became 'chemical-free')

Field no. (see map)	Field name	First year chemical-free	Total acreage	1984	1985	1986	Acres per crop	Remarks
1	Coombe	1981	48	LL4 WW1	LL5 WW2	SW LL1	18 30	
2	Lower Barracks	1981	7	LL4	LL5	SW	7	
3	Barracks	1985	31.5	WW2	LL1	LL2	31.5	
4	Above Charlton Barn	1984	36	LL1	LL2	LL3	36	Plough for WW
5	Below Charlton Barn	1980	40.5	WW1	WW2	LL1	40.5	
6	Above Cat	1985	38	SB	LL1	LL2	38	
7–8	Milestone	1984	52.5	LL1	LL2	LL3	52.5	Plough for WW after hay
9	Lower Waights	1984	14	LL1	LL2	LL3	14	Plough for WW after sheep
10	Upper Waights	1984	32	LL1	LL2	LL3	32	Plough for WW after sheep
11	Six Acres	1985	20.5	SB	LL1	LL2	20.5	
12	Ten Acres	1985	11.5	SB	LL1	LL2	11.5	
14	Under Road	1972	43	Clover	WW	WO	43	U/S LL
15	Baldhams	1981	25	LL4	WW1	WW2	25	U/S SL
16	Charlton North Field West	1976	45	LL4	WW1	WW2	45	U/S SL
17	Charlton North Field East	1976	37	WW1	WW2	R/Cl.	37	Plough for WW after hay
18	Haystock	1983	12	LL2	LL3	WW1	12	
19	Rushall North Field West	1971	30	WW2	LL1	LL2	30	
20	Rushall North Field East	1972	31	LL1	LL2	LL3	31	Plough for WW
	Ducks Acre	1972	14	WW1	WW2	R/Cl.	14	Plough for WW
21	Post Office	1978	13.5	SO	LL1	LL2	13.5	
22	Cuckoo Pen	1978	20.5	LL2	LL3	WW1	20.5	
23	Well	1978	33	LL2 LL2	LL3 LL3	WW1 LL4	16 17	Plough for WW after hay

No.	Field	Year	Acres				Acres	Notes
24	Plough	1983	44.5	LL2	LL3	WW1	44.5	Plough for WW after hay
26	Pump House	1978	35	LL2	LL3	WW1	13.5	
				LL2	LL3	LL4	21.5	
27	Dumplings	1980	30	WW1	WW2	LL1	30	
28	Bellows Piece	1984	14.5	LL1	LL2	LL3	14.5	
30	Rushall Barn	1983	30	LL2	LL3	WW1	30	Plough for WW after sheep
31	Reservoir	1972	27	Clover	WW1	Roots	27	Swedes, mangolds, kale
32	Parkway	1981	21.5	LL4	WW1	WW2	21.5	
33	Sargents Barn East	1983	23.5	LL2	LL3	WW1	23.5	
	Sargents Barn West	1986	24	Roots	SB	LL1	24	
34.	Jopping Hole	1980	19	WW1	WW2	LL1	19	
35.	Above Lawns	1970	19.5	WW2	Clover	WW1	19.5	
36	Under Lawns	1975	21	WW2	Clover	WW1	21	
37	Rushall Drove West	1973	40.5	WW1	WW2	R/Cl.	20	Plough for WW after hay
	Rushall Drove East			LL2	LL3	WW1	20.5	
38	Garston Corner	1973	36	WW1	WO	LL1	36	
39	Overbrow	1982	30.5	LL3	LL4	WW1	30.5	
40	Alexanders Cleeve	1982	44	LL3	LL4	WW1	44	
41	Searchlight	1981	55.5	WW1	WW2	LL1	30	U/S SL
				LL4	WW1	WW2	25.5	
43	Upavon Side	1986	40	Roots	SB	LL1	40	
44	Flagstaff	1980	49	WW2	Roots	SW	49	
45	Woodbridge	1979	16	SO	LL1	LL2	16	
47	Black Knoll	1979	30	WW2	SB	LL1	30	
48	Little Garston	1970	18	Beans	SW	LL1	18	

LL = long ley; R/Cl. = ryegrass/red clover/short ley; SB = spring barley; SL = short ley; SO = spring oats; SW = spring wheat; U/S = undersow; WO = winter oats; WW = winter wheat. The number after the letters indicates which year the crop is in (for example) LL2 means second year in long ley. For the layout of the farm, see the map on pp. 4–5.

feeling our way at that stage, as ADAS and the advisers from the fertilizer companies didn't want to know): if the whole farm was to be converted it was no good just starting with the best land, only to find out too late that we couldn't cope with the poor land. So these two fields were scheduled for conversion. Little Garston was planted to winter wheat on 10 October 1969, and our method then was simply to cut off all fertilizer and sprays. That crop of wheat was poor and to compound the felony we planted winter wheat again for the 1971 harvest! So it is hardly surprising that in the autumn of 1971, as I see from a note in my cropping book, the field was dirty and required extensive cultivations to kill cooch grass and weeds. To salve our consciences, I suppose, we dunged it after cleaning and planted spring wheat in March 1972. This was undersown to a grass ley (a process explained under 'April' in chapter 3).

This little sequence reveals nearly all the mistakes in the book! In the first place, two winter wheat crops on land just deprived of fertilizer is asking for trouble. We made the mistake, common to many would-be organic farmers, of equating 'no chemicals' with 'organic'. We just stopped applying chemicals and hoped that natural growth would take over. Hopeless! It takes several years for land to regain its optimum population of soil micro-organisms and, until that level has been achieved, output will be limited. From this it follows that the first requirement when converting a field from a chemical system to an organic one is to build up the soil population as quickly as possible — which, given that it has been estimated that in a saltspoonful of fertile earth there are more living organisms than there are people in the world, is no mean task. We have found that it takes some five or six years on the system we adopted for land to regain its full potential fertility (a direct reflection of the level of the soil population), but with the benefit of hindsight and the experience of field 23 (see below) this period could well be reduced by introducing dung at an early stage of the conversion. It appears that a chemical regime, with its high input of fertilizer and spray, depresses the vital soil population, which would account for our poor yields during the first few years after conversion, and that it is not until this population has been built up that reasonable results can be achieved. To my mind the whole question of the soil population is of the utmost importance, and, as John Wibberley of the Royal

Broadcasting grass seed into winter wheat in April (undersowing)

Agricultural College, Cirencester, concluded in his article 'Micro-biological perspectives in soil management' in *Soil and Water*, the journal of the Soil and Water Management Association, 'A long-term appreciation of the roles of soil micro-organisms is needed for effective farming, and as farmers we must call for and support continued relevant research in this relatively neglected field.' Because Wibberley's article provides such an admirable general introduction to the subject and is in large measure accessible to the layman as well as to the specialist, substantial extracts from it are reprinted in Appendix D, by kind permission of the Soil and Water Management Association.

Secondly, there is absolutely no need to saddle yourself with a big handicap at the outset by experimenting with a field that is dirty when with a little foresight you can start with a clean field. Another fundamental mistake was to leave the field fallowed for the winter of 1971—2 after the dung had been ploughed in. It is essential, in this system, to have something growing over the winter, even if it is only weeds. By leaving land in bare fallow throughout the winter months you may lose many of the nutrients present in the soil. The aim is to have a green cover, be it weeds or crop, that will lock up these nutrients. If this is done with weeds, ploughing in the spring will release the nutrients for the chosen crop then planted. When a crop is sown, of course, the nutrients are utilized straightaway.

In field 35, Above Lawns, we tried a very different technique. At least it was a positive effort to do things correctly, even if the results were scarcely any better than with Little Garston, where we just blundered on and hoped for the best. Above Lawns, it will be remembered, was the worst field on the farm, and in an effort to do something about it we turned to a book called *Humus and the Farmer* by Friend Sykes, published in 1946. In chapter 5, 'Making a new pasture', Sykes describes his method of establishing a ley, and we followed this exactly. After winter wheat in 1969 the field was ploughed during the winter. Then it was ploughed again three times before working it down to a seed bed by the end of June. On 1 July, mustard was planted at the rate of 20 lbs per acre. This was ploughed in on 31 July, the land worked down and rye drilled on 1 August. On 2 August, grass seeds were broadcast and rolled in. According to the book, all we had then to do was to graze the rye and grass seeds with sheep and bullocks, and by the following summer we should have one of the finest swards of grasses and clovers that we had ever seen. Unfortunately, things didn't turn out quite like that, and the note in my cropping book for 1971 reads 'Poor'! As we now know, we committed several cardinal sins against good organic practice. We left the field in fallow over winter, allowing nutrients to leach away. We ploughed three times during spring and early summer, which had the effect of wasting what nitrogen there was. Cultivations during the warmer months promote the mineralization

Loading the dung spreader with a rough terrain fork lift

of organic nitrogen, and if no crop is present to take it up it is lost. This is a criticism that scientists often level at us relative to 'bastard fallows' (see Glossary and later in this chapter) that we adopt after a ley to prepare a good clean seed-bed for the wheat. They tell us that by adopting a 'bastard fallow' technique we lose many of the nutrients that should be kept for the wheat, but we feel it is a price we have to pay to ensure a clean farm. And I am not convinced that they are right anyway! I feel that the mineralization of nitrogen from the decaying turf only takes place in warm, damp conditions, and these seldom occur in August and September. We have tried leaving a ryegrass/red clover ley to regrow after taking one crop of hay, grazing the sheep on the aftermath (see Glossary) and then breaking it up — but with not very satisfactory results. So we have now reverted to breaking it up as soon as possible after the hay has been cleared to make sure we have a clean field for the wheat crops.

In 1971 we scheduled field 31 — Reservoir Field. This is one of the highest fields on the farm, and we thought we had better investigate whether the land on the high chalk was capable of growing a decent crop without chemicals. With our experience of Little Garston and Above Lawns we decided to establish the ley by undersowing a crop of spring barley with the seed mixture we use for the purpose (see table 6.3, p. 95). The barley was planted with fertilizer and sprayed as necessary, but after that the field was 'chemical-free'. Beginning in 1971, with one or two exceptions that proved pretty disastrous, we adopted this technique on all our fields and it served us very well. After three years in grass the field was deemed to be free of all residues and was allowed to produce wheat for the flour mill. Reservoir Field, in spite of its height, has always yielded well, and we have had several crops of wheat yielding over 2 tons per acre.

Another interesting field is Well Field — no. 23. This was undersown in the spring of 1977 and completed its three years 'probation' in 1980, when it was planted to winter wheat. My cropping book notes reveal that it yielded 2 tons per acre, that mixed

Undersown red clover growing up through wheat stubble

in with the wheat were a lot of grass weeds and that because of this we burnt the straw, something we haven't done since. A second winter wheat followed in 1982 and this was undersown to a long ley. For some reason that I cannot recall, we dunged this field in the autumn of 1982 — probably because we had nowhere else to put the dung! (We are not always as scientific as we might be!). Be that as it may, the field has never looked back. That ley, now coming into its fourth year, has always produced a lot of grass. It starts into growth earlier in the spring than most of our other leys and continues all through the summer and well into the autumn. The only significant difference between this ley and all the others, so far as I can tell, is that we dunged it during its first autumn. Maybe this is what we should always have done. Maybe the dung allowed the soil population to build up much more quickly than would otherwise have been the case. But this is conjecture: there are so many facets of organic farming which we cannot hope to understand with our limited resources and knowledge, and what is needed (as already mentioned and as discussed more fully in chapter 11) is for the Ministry of Agriculture to accept its responsibility to carry out careful long-term research into organic farming methods.

Each year from 1970 onwards we scheduled one or more fields to become 'chemical free', and in 1985 the last two fields were converted. As already noted, conversion has mostly been carried out by way of undersowing a spring barley crop, having first made sure that the field was as clean as possible. If the field had a lot of couch grass we did use Round Up and similar sprays, as it seemed rather pointless to leave the cleaning operation until after the field had been scheduled. But, once it was in the system, of course, sprays were out and we had to rely on rotations and cultivations to keep our fields clean. (See chapter 6.)

Mention has been made of techniques that proved rather disastrous. I am thinking particularly of fields 19 and 37, Rushall North Field West and Rushall Drove. In both cases we sowed to a second long ley — i.e. after five or six years in the organic system, as can be seen from table 5.2.

A partially cut field of grass for hay

Table 5.2 Rushall North West and Rushall Drove, cropping notes 1981—6

Year	Rushall North Field West (converted 1971)	Rushall Drove East (converted 1973)
1981	Roots grazed by sheep	Roots grazed by sheep
1982	Spring barley	Maris Wigeon winter wheat; ploughed after harvest and grass seeds planted
1983	Rye, grazed by sheep	Long ley 1 — poor
1984	Avalon winter wheat; grass seeds broadcast onto stubble	Long ley 2 — poor
1985	Long ley 1; dunged in May	Long ley 3 — better
1986	Long ley 2 — good	Brimstone winter wheat — good

In the case of Rushall Drove we tried to reseed directly after a winter wheat crop had been cut. We ploughed the field and worked it down to a seed-bed, broadcast the seeds and rolled them in. With Rushall North Field West, the reseeding again followed winter wheat but in this case we found a beautiful tilth in the stubble — about three quarters of an inch of friable soil — so decided to broadcast the seeds, harrow and roll. In both fields the seed germinated very well and then seemed to go into suspended animation. Both were very slow to start into growth in the spring, and in fact, produced nothing until well into the summer. With the experience of Rushall Drove behind us (1983), we applied a light coating of dung to Rushall North Field West in early summer 1985 and from then on we had a good growth.

These two fields illustrate, I believe, the limitations of organic farming. You cannot expect your land to produce more than its potential, and anyone brought up on the chemical system, as I was, sometimes finds it difficult to appreciate this. These two fields had been exhausted by the winter wheat and there was little or no free nitrogen left to give the seeds a good start, so it was not until the clover in the one case, and the dung in the other, made some nitrogen available that the leys produced anything worthwhile. In fact, the field without the dung did not produce a good growth until its third year, by which time the clover was making a real contribution.

Little mention has been made so far of the stock. Obviously, with this system of farming there is a lot of grass and this is utilized by the beef cattle and sheep, with a little help from the horses. We started by grazing new leys with the sheep in the early part of the grazing season, but because our growth is slow and late (until the soil population of micro-organisms becomes active in the warmer weather we get very little) they tended to eat out the new leys, which then had a job to recover. Ideally we like to use the lighter bullocks, as they don't graze so tightly, but if we do use the sheep we try to move them on before they do too much damage. This is another limitation of an organic system — you do have to wait for the weather to warm up the soil, as you cannot liven things up with a bag or so of nitrogen. The month of May does try our patience when we see our neighbours' fields growing away like mad, and ours still bare and seemingly asleep.

I have frequently been asked what is the best way to convert to an organic system from a chemical regime. Obviously, each farm presents its own specific problems and no one programme can be suitable for all farms. For instance, the enterprises will differ — some will be dairy farms, some mixed, some arable and some specialist vegetable-growers. I can only suggest a way that I think would work on most mixed farms, but, over and above the mechanics of conversion, there are one or two fundamental principles that anyone thinking of converting should carefully study.

The first is the motive of the farmer himself — does he want to 'go organic' to make quick profits or does he really feel that this is the best way forward for him? If the former, then he is barking up the wrong tree: seeing how long the conversion process takes, a quick return is out of the question, and, with things as they stand, conventional farming is more profitable anyway. This is not to say that one cannot expect a good profit from organic farming: rather, the point is that, without a firm belief that organic methods are environmentally and agriculturally preferable — that they offer the best hope of a sustainable system of agriculture over the long term —

the dedication that organic farming requires will be lacking and one is unlikely to make a success of it.

Secondly, there is a need for careful planning. Do you have a market for your produce? Remember, your output will be less than when you used chemicals, so it is essential to do some research to establish just who and what your market will be. Then, are you prepared to suffer a drop in income for the years taken to convert your farm? We have found that it takes five or six years for the soil to regain its potential after having been on a chemical regime for many years, so not only is the productivity of a well established field farmed by organic methods less than that of the same field under a chemical regime, but there is also a large initial drop in productivity at the beginning of the conversion process, followed by a gradual rise as the field recovers. This must be realized and allowance made for it when budgeting.

Having satisfied yourself that you can overcome all these problems, you must formulate a plan to convert the whole farm. As I mentioned earlier, we gave ourselves twenty years to convert our 1650 acres. We actually did it in fifteen, and I now believe it is practicable to think in terms of ten years — one tenth of your acreage to be converted each year. Obviously, you can't be exact as to the yearly acreage, as it will vary slightly with field size, but one tenth per year is a good target. My experience is limited to a mixed stock and arable farm, so what I offer now are what appear to me to be the important points to remember when converting such a farm.

It is essential to start with a clean field. If there is couch grass present, then kill it. You can still use chemicals at this stage, so make certain that your chosen field is clean. This can be done with one of the proprietary couch-killers after harvest the previous year, and time can be allowed for it to act as you will not want to plant the field until March or April, when spring barley should be grown in the conventional way, using fertilizer and sprays as necessary. This crop should be undersown with a grass-seeds mixture, the actual constituents being determined by your locality, but a broad range of

Hereford × Friesian steers in front of Rushall Church

varieties as described in chapter 6 should be used. After harvest, the grass seeds will start growing away and, from this time on, no more chemicals, fertilizer or spray, may be used. With the benefit of hindsight, I would now recommend that these grass seeds should receive a top dressing of well-rotted dung during September or early October. This will enable the soil population of micro-organisms to build up very much more rapidly than if left to their own devices. Should the autumn be open and mild, it may be necessary to run the sheep over the field before Christmas to trim off any growth above 3—4 inches. Not only will this tidy up the field for the winter; it will also have the effect of firming up the soil round the roots of the grass.

The following spring, if dung has been used, there should be a reasonable growth. If no dung was used, growth will be slow and poor. Young bullocks, not too heavy, are the best for grazing during the first spring and early summer, as they do not graze as tightly as sheep. Sheep are preferred as soon as the grass is really established. The field stays in grass for three years before being ploughed up to grow wheat. We cut for hay in the third year and then immediately break the field up with a chisel plough (see Glossary). The object is to kill off the turf, which we hope is dense by this time, as any grass not killed will appear in the following wheat crop — no sprays allowed! These operations following the hay constitute what is called a 'bastard fallow' and it ensures that not only the turf, but also any couch that may be in the field, is killed. Turf and couch can be killed in hot dry weather quite easily — it is not so easy in a very wet summer as in 1985, but a good measure of control can be achieved if you keep at it whenever the weather allows.

The hope is that the field will be clean by the beginning of September, when it should be ploughed and left for a week or so before working it down to a seed-bed by the beginning of October. This is when your control over annuals should be achieved. The soil always has thousands of weed seeds in it, and they germinate when they find themselves in the top 2 inches or so of the soil. Below about 3 inches they remain dormant, so it is important when practising a 'weed strike' to avoid any cultivations below 2 inches. The seed-bed, called sometimes a false seed-bed, is left from the beginning of October until drilling starts, some time after the middle of October, during which time the weed seeds in the top 2 inches will germinate. They will in all probability still be below the surface of the ground after this interval of about two weeks, but at this white-string stage they are very vulnerable and the mechanical actions of drilling — the drill itself and the harrows that follow — will eliminate the majority and leave you with a clean crop.

I have gone into some detail about the bastard fallow and the weed strike (false seed-bed) techniques as they are the basis of an organic farmer's efforts to keep his farm clean (see further in chapter 6). Another helpful control measure is to make sure that weeds in the leys are cut before they have time to flower and set their seeds. If cut when they are just showing colour in their flower buds, the plants will not put up another flower head and will die.

By this time, year four of the ten you are taking to convert, several other fields will be in the system and it is a time of great temptation. That part of the farm that is still receiving chemicals will be producing the usual 4 tonnes per acre of wheat or barley, while the first field to be converted to organic methods (the others still being in grass), will be producing what looks, comparatively, a very thin crop of 2 tonnes per acre if you have done everything right, less if things have gone wrong. You will have doubts as to the viability of the system, and you will probably fall into the trap of thinking that a crop of 2 tonnes per acre cannot possibly pay its way. This is the time when you will really discover if you are an organic farmer! It is the time to remember Kipling's poignant lines

> If you can trust yourself when all men doubt you
> And make allowance for their doubting too. . . .

Provided you have a reasonable crop you will find yourself gradually becoming more convinced that what you are doing is right. Your fertilizer and spray bill will be diminishing, you will be proudly selling your produce to willing buyers with no help from the taxpayer, and your delight at producing a good healthy crop without the aid of the chemist will gradually overcome your doubts. It *is* a critical time, one that the aspiring organic farmer must anticipate

and prepare for, but it quickly passes as more of the farm becomes organic and the profits continue to hold up. At the end of the ten-year conversion period you will have a farm on which it is a pleasure to work and that is a joy to behold.

All this applies to the owner occupier (I am one), and it must be admitted that there are difficulties for the tenant farmer. For a start, he must convince his landlord that the organic system is a reasonable one to adopt — many landlords still feel that if any of their tenants take up organic farming their land will quickly become derelict. While this is certainly not so, it only needs a few mistakes by a tenant for the landlord or his agent to become convinced that his worst fears are about to come true.

Perhaps the most difficult question the tenant has to face is the one of succession. Suppose it takes him ten years to convert his land and he has a son or other relation to follow him, he cannot be certain that that successor will be allowed to have the tenancy. As the conversion of a farm is such a long-term project, a tenant may well feel he is not justified in undertaking it unless he knows that his efforts will not be wasted by his landlord letting his farm to a chemical farmer when he gives up. These are real worries and I would recommend that both parties, tenant and the landlord, should be in accord on the subject before any conversion is undertaken. In the long term I would hope that an organic farm which has been allowed to sell its products under the Soil Association symbol will be a saleable asset, in the same way as a farm with a healthy milk quota is, so the landlord would benefit by keeping it organic. But that may still be a long way away. In the meantime, I would suggest that the tenant consult with his landlord very carefully before embarking on what will be, in effect, an act of faith to posterity.

One of our long leys just before cutting for hay

6

Organic husbandry

Everyone can tell you what a farmer is — one who tills the soil, rears stock, and so on — but does everyone really know what farming means? Up to the last war there was not much doubt, as all farmers adopted the system of farming that suited their district, and which had been proved over many years to be sustainable. Thus you found the grassy western half of Britain supporting cattle for milk and beef, with a little corn to feed these animals, while in the east corn-growing was stronger, with yarded animals to provide the dung to go on the land. Although the emphasis was different, all farms were 'mixed' and the system of farming 'sustainable'.

It is no longer that simple. Many farms in the east — and some, regrettably, further west — are now stockless. Instead of recycling their wastes, notably straw, through animals to convert it into a natural fertilizer, they buy in tons of chemical fertilizers, which is much less trouble, and grow corn continuously. Then, when the corn grows very lush and disease sets in, it has to be sprayed perhaps six to eight times a season. And the ultimate irony is to spray the corn with a growth regulator to prevent it from growing as tall as the fertilizer it has been fed with would have it grow. This system has been positively encouraged by successive governments, and so little blame other than short-sightedness should be laid at the farmers' door. But the system *is not farming*. Take away the artificial inputs, which are largely derived from (not unlimited) fossil fuels, and farms run in this manner would be uneconomic due to the deterioration of their soils and the depletion of the soil populations.

Haymaking in the park with Rushall Church in the background

All farming, to my mind, should be as near sustainable as possible. Systems which ignore nature and impose an unnatural chemical regime on a farm are not, I submit, farming. They are far more akin to exploitation or, to use a business phrase, asset-stripping. Asset stripping in the business world has, I understand, been banned. When will the thoughtless exploitation of our farmland — and our farm stock — be similarly banned?

I set out in this chapter to describe the main features of organic husbandry. I make no claim to know everything about the subject, but there are four questions about organic farming that I am always being asked and that I shall do my best to answer. The questions are:

1 How do you control weeds?
2 What rotations do you use?
3 What seed mixture do you use?
4 Does it pay?

The last of these questions will be dealt with separately in chapter 9, with particular reference to cereal farming, as it is really the key to the whole subject. The others concern aspects of organic husbandry and will be answered here in the above order.

Controlling weeds

In our experience, most people expect to see our cereal crops smothered in weeds, and they do perhaps have good reason for expecting this when they look around the countryside in early summer and see the strips left by the sprayers on conventional farms. In spite of being sprayed regularly for many years, these strips will be full of weeds — charlock, poppies, wild oats, sterile brome, or a mixture of all of them. So the expectation is that the organic farmer, the one who uses no sprays at all, will have all his cornfields looking like these strips. This picture owes something too to the publicity and advertising put out by the agro-chemical companies, and here I must pay reluctant tribute to their skill. Farmers, generally speaking, have been completely duped into believing that it is impossible to grow a good clean crop without their help. Just consider for a moment some of the names these companies have dreamed up to help sell their poisons: Assassin, Avenge, Commando, Crusader, Harrier, Hotspur, Missile, Musketeer, Patrol, Radar, Round Up, Swipe, Tilt, — and there are many more. Tilt must be my favourite name, as it conjures up a picture of a knight in shining armour galloping to the rescue of some dazzling maiden, but all these names are designed to convey a macho image of good overcoming evil.

The organic farmer, however, has to rely on other methods. In my view, the secret of effective weed control without sprays lies in four basic principles, all very simple in themselves, but together embracing the whole concept of sustainable farming.

1 Not allowing annuals to seed.
2 Timeliness of operations.
3 Constant attention to detail.
4 Working with, rather than against, nature.

Quite obviously, not allowing annuals to seed means that in a cereal crop there must be no weeds to go to seed, as any that are present will have flowered and seeded by harvest time, and hoeing cereals on a large acreage is just not practical. The method we use to prevent seeding is the 'weed strike' or 'false seed-bed' technique, which, as briefly outlined in the preceding chapter, involves preparing the ground for planting some ten to fourteen days before drilling actually takes place. With normal weather conditions and a little bit of luck, all the weed seeds in the top 2 inches or so will start into growth, ideally to the white-string stage, when they are very vulnerable, and the subsequent mechanical actions of drilling and harrowing will eliminate the majority of the weeds and volunteer cereals. The number of weeds that survive this process is usually too small to depress the crop's yield. The effectiveness of the technique is borne out by the preliminary results of an ADAS monitoring project recently undertaken at Rushall (on this project, see chapters 9 and 11). Some, and I would not disagree with them, say that a certain level of weeds is actually advantageous to the ecology of a field, and, if this is so, then perhaps the converse is true — that a completely 'clean' crop may be harmful to the environment. With leys, of course, preventing annuals from seeding means topping with a mower just before the seeds are set — too soon and the plant will put up another flower, too late and the seeds are viable whether cut or not.

This leads on naturally to my second principle, timeliness of operations, which in my opinion is one of the most critical factors in the running of an organic farm. I think most farmers would agree that, provided you don't miss an opportunity in farming, nature will atone for her aberrations. On the other hand, if you don't take advantage of the opportunities offered, you will in all probability lose out. This applies especially in organic farming, where you

A ryegrass/red clover mixture ready for the mower

cannot call in the chemist to rectify your mistakes. By timeliness I mean such things as waiting for a weed strike; ploughing for spring crops in February or March, rather than November or December, delaying the start of autumn sowing until mid October; and cutting thistles in July:

> Cut 'em in May, they'll grow next day,
> Cut 'em in June is a month too soon,
> But cut in July, they're sure to die.

The date of drilling autumn-sown crops is important. By not drilling till mid October you achieve a very good control over blackgrass, which can be a serious problem on our heavier land. This is a far cheaper control measure than planting early and then spraying. As my farm manager puts it, 'If you plant in September, you must spray in November.'

So far I have dealt with the annual weeds, but in farming the perennials too can be a real problem. The worst is, I suppose, couch of both sorts — what in Wiltshire are called running couch (*Agropyrou repens*) and knotty couch (*Arrehatherum elatious*) — followed by docks, nettles and thistles. With no sprays to help us we rely on the bastard fallow to control any build-up of these problems. In a field's last year as a ley we cut hay and then immediately break the ground up. Thus we have what are normally the three driest months — July, August and September — to work the ground to kill off the ley and any perennial weeds that might be there. This is really the only opportunity we have of doing this and it is vital to the maintenance of clean fields. The experts tell us that we lose a certain amount of organic nitrogen every time we cultivate a field in this way, but we feel it is a price we have to pay to maintain the cleanliness of the farm.

These are the basic ways of growing weed-free crops, but they can be helped along by one or two other devices. It is fairly obvious that a tall crop has a greater smothering effect than a short one — it is easier to grow a clean crop of kale than a clean crop of swedes, for instance — so we try to use tall varieties of wheat. Unfortunately, modern plant-breeders seem to find it advantageous to breed dwarf varieties — presumably because these are more likely to remain standing at harvest after having been treated with as much bag nitrogen as the farmer can afford — so the newer varieties are not as good for the organic system as the older, taller ones. Two other dodges help to achieve a smothering effect: a narrow-spaced drill and a slightly higher seed rate. We have tried both and they can be a help, but can never replace the basic principle: destroy the weed seedlings before you plant your crop.

By 'constant attention to detail' (my third principle) I really mean the farmer's presence. The old adage that the best fertilizer is the farmer's foot is so very true of organic farming. For instance, the timeliness mentioned above is really only attainable if the farmer is on hand to make the decision. He must live his farming — be a 'gut' farmer if you like — and have that feel for his crops and his stock that was possessed by the farmers of old. They knew their fields, they knew their stock and they knew there were no short cuts. Instead of carrying tape-recorders and disease identification charts, as modern farmers now do, they would carry a hand tool as they walked their fields. It is surprising how many thistles can be destroyed in the spring and early summer by using a thistle spud, and it is very relaxing and satisfying to hook them out as you go about your business and to know that they will not grow again. The trick is always to have your thistle spud with you (otherwise you get frustrated when you come across thistles you can't remove), and to make sure you cut them off low enough. If you merely slice them off at ground level they will send up lots of side shoots like a 'stopped' chrysanthemum, so you have to make sure the cut surface is about an inch below ground and white — if it shows green it is not low enough. Later in the season, a long-handled hook is the tool to use to deal with any odd nettles or thistles that have managed to survive. While the total number of weeds destroyed in this way may be small compared to what can be achieved with the broad sweep of a sprayer, the method is very therapeutic for the farmer and is also very selective — if there is a nest in a bunch of nettles it can be seen in time and the eggs left to hatch. Now, unfortunately, the chemist can cover up the results of bad husbandry practices and only the accountant realizes the true cost of a clean crop. It takes a true farmer to farm without chemicals.

This 'feel' for one's farm is really the basis of the last of my four principles: work with, rather than against, nature. Too many farmers — not all of them the brash entrepreneurial types usually associated with the absentee, institutional farming company — rely on the scientist to solve the problems that they, the farmers, create in their search for ever higher production. This was brought home to me one day when talking to a friend who leads the field in his efficiency as a 'chemical' farmer. When I asked him whether he wasn't worried at the increasing use of chemicals to combat the seemingly endless succession of new diseases, he replied that the scientists had always kept one step ahead and he expected them to continue to do so. A somewhat precarious philosophy — fine while it lasts. The agro-chemical companies obviously encourage this belief, as it makes the farmer ever more dependent on them. Indeed, the cynic who said that this year's spray contains next year's disease had a point!

In the extreme case, it could be said that a farmer who relies on chemicals uses the soil of his fields merely as an anchor for his crops which he feeds with chemicals. The organic farmer seeks to obtain the optimum from any given field and this varies with every field. This optimum can only be achieved by feeding the soil so that the myriads of organisms in the soil can perform their function efficiently. The two philosophies are so far apart that it takes a very determined and thoughtful farmer to embark on the conversion of his farm.

Crops are, of course, not the only aspect of farming. Since, in my view, all organic farms should have stock, animal husbandry is also very important. Quite apart from the fact that the general public is becoming restive at the way some farm animals are treated, the philosophy of the organic farmer would not permit some of the practices common today. Caging hens in batteries, penning calves in crates, rearing table birds in broiler houses, implanting beef cattle with hormones, prophylactic treatment of pigs with antibiotics — all these barbarous modern techniques are definitely working against nature. So long as these practices remain legal, farmers will go on using them, which raises the question of whether nature will fight back. Indeed, there is a growing feeling among farmers and scientists that one day we shall suffer a disease disaster, in our crops or in our stock or in both, that we shall be utterly powerless to control. For instance, it is said that scientists are beginning to worry about the

A swathe of freshly cut grass for hay

long-term effects of an apparently benign chemical on a seemingly unaffected species. Consider the following hypothetical case: an oak tree in a wood adjacent to an arable field that is sprayed once or twice a year with a herbicide might show no ill effects for five, ten or even fifteen years, but then suddenly dies, succumbing at last to the debilitating effects of that constant bombardment, year after year, with a chemical that no one had even dreamed might affect it. Is this sort of consequence so unlikely? Already we have two diseases of barley — barley yellow dwarf virus and barley yellow mosaic virus — that are untreatable when once established. The dwarf virus is spread by aphids and the only control is to attack the disease vector, in this case the aphid. In the case of the mosaic virus, which is a soil-borne disease, control depends on using varieties that have a known tolerance to the virus. Once established, both of these diseases, for which there is no chemical control, can decimate a crop of barley, so it is not being too fanciful to suggest that there may well come a time, if the use of chemicals continues to expand, when a new and untreatable disease strikes the nation's wheat crops, causing bankruptcies in farming and food shortages in the shops.

Rotations

Our original rotation, once we had established it, was very simple: three years ley followed by two years winter wheat, the second undersown to another long ley. This worked very well and we found that we could grow two good wheat crops in this way. But it is obvious that this rotation allowed us to have only two fifths of the farm growing corn in any one year, and, with the ever-growing demand for organic wheat, this was not good enough. Our forefathers used to manage their land so that about half of it was planted to corn. This they achieved by using the old four-course rotation, which in Wiltshire used to run something like this:

Year 1   Wheat
Year 2   Barley, half field undersown with ryegrass/red clover
Year 3   Roots, folded by sheep; new field hay
Year 4   Roots or vetches, folded by sheep; old field hay, bastard fallow

And then back to wheat.

Top left: Wheat

Top right: Barley

Left: Oats

We have been trying to adapt this to suit our long-ley system, and what we have come up with is this:

Year 1   Long ley, grazed by cattle and sheep
Year 2   Long ley, grazed by cattle and sheep
Year 3   Long ley, hay followed by bastard fallow
Year 4   Winter wheat
Year 5   Winter wheat, undersown with ryegrass/red clover
Year 6   Hay, aftermath grazed by sheep, then ploughed
Year 7   Winter wheat
Year 8   Winter oats

As every farmer knows, you can't farm by the book! The weather has the final say in what happens every season, and, while you may try to stick to a rotation, the weather may decree otherwise. So far we have only been trying the above eight-year rotation for two seasons, both of which have been somewhat atypical. The 1985 harvest was very wet and followed a very severe winter when we lost a lot of plants on the higher ground. And the red clover kept on growing until you could not see the wheat for the red flowers! These were climatic problems and we have resolved to plant the ryegrass and red clover much later in future. But the real lesson we have learnt from the two years is that a one-year red-clover ley does not provide enough nitrogen for two winter corn crops. In 1985 the first crop of winter oats on this system only yielded about 1 tonne per acre, and after the severe frosts in February 1986 the second crop of winter oats had to be redrilled with spring oats. To overcome this and to try and retain the four wheat crops in the eight years we are proposing to undersow the winter wheat crop in year 7 with straight red clover. This will be grazed by the sheep in the autumn and left over the winter before ploughing in February/March for a spring crop — either wheat or oats, as we need about 100 tonnes of oats for the horses and calves. In this way we hope to have enough available nitrogen in year 8 to grow a decent crop of corn and start the rotation all over again.

This is an eight-year rotation and I am getting no younger! What a pity that the Ministry of Agriculture refuses to undertake the necessary research so that the ever-increasing number of farmers who are thinking of converting to organic methods would not be so

Ploughing wheat stubble for a second winter wheat crop

dependent on trial and error in establishing a suitable rotation. (See chapter 11.)

Experience so far suggests that we may have to adopt two different rotations for the heavy and the light land. I suppose that this is pretty obvious to the objective observer, but we have been trying to establish just the one to make life easier. Soils do vary in their optimum productivity, and to expect a light, thin, chalky soil to perform as well as the heavier soils of the vale land is perhaps rather optimistic. I foresee an eight-year rotation on the heavy land and the original five-year rotation on the hillsides as offering possibly the most sustainable system for the future.

No mention has been made of roots in the rotation, largely because we grow only about 60 acres a year. As any hunting or shooting farmer will tell you, these root fields always seem to be where the foxes and pheasants would like them to be and not necessarily where the shepherd or the farmer would have them! Be that as it may, the kale is grazed after the shooting season by the sheep, and the clean swedes that we grow are fed chopped with rolled oats to the calves (old-fashioned maybe, but nevertheless excellent) and whole to the outliers. The sheep clear up. We have been tending to reserve the dung for the root acres (we have enough for about 60 acres a year), but, as mentioned previously, we should perhaps have concentrated on the new leys, to try to boost their soil populations earlier in the conversion period and so cut the time taken to achieve their optimum.

Seed mixtures

Given that the long leys are really the basis of our system, its effectiveness therefore depends on the seed mixture that is used. Early on in the exercise I thought I had better seek expert guidance on what would be the best mixture for our purposes, so I called up a friend of mine who ran a well-known company of agricultural merchants. As it so happened, his firm had just been taken over by an international group which had appointed two grass-seeds experts to his old company, so he suggested he should bring both of them along with their recommendations.

Not being a scientist myself, I asked our local veterinary surgeon if he could tell me the requirements of the grazing animal, by way of main elements and trace elements, for full health. He had to do some research, but was able to provide the information quickly enough for me to have it in front of me when my friend and his two experts arrived. Looking back, I think they had felt that mine was a routine enquiry and that they could just jot down some suggestions on the back of an envelope as they came along. But armed with the vet's information I began questioning them as to the source of the various trace elements — which herb (in the wide sense) would provide which trace element and how to balance the whole mixture. They looked at me as if I was someone from outer space, so I rubbed it in by telling them that we required a balanced mixture that would keep our animals healthy without recourse to a lot of supplements. I remember waxing quite eloquent on the subject and telling them I was not interested in any of their mixtures designed to produce tons of wet grass per acre with the help of tons of fertilizer (also sold by their company!). The upshot was that I sent them away with a flea in their ear and told them that if they wanted to supply me with grass seeds they had better do their homework and balance the requirements of my stock against their mixture. Throughout all this my friend said nothing — he just sat in the corner smiling quietly to himself.

About a month later I had a letter from the experts. To give them their due, they really had done their homework — they had found out exactly what each variety of grazing animal required and which grass or herb supplied it — and they had prepared a mixture (see below) that satisfied these requirements. To be fair to them I have included their explanation of how the mixture was arrived at, but the most interesting part of this story to me lay in their statement that they had found the whole exercise most interesting as they had never been asked that question before. Surely this is a very sad reflection on agricultural teaching and the state of farming generally!

Included with their letter were lists showing the mineral content of herbs and grasses (table 6.1), and the trace-element requirements of livestock (table 6.2). These are reproduced here not only because they contain the data from which our balanced mixture was built up, but also so that interested farmers can check whether their ley

Table 6.1 Mineral content of herbs and grasses

Species	P (%)	K (%)	Mg (%)	Cl (%)	Na (%)	Fe (p.p.m.)	Mn (p.p.m.)	Cu (p.p.m.)	Co (p.p.m.)
Yarrow	0.91	4.42	0.98	0.53	0.06	294	48.03	10.06	0.17
Burnett	0.62	2.22	1.82	0.15	0.07	249	31.02	8.00	0.18
Plantain	0.75	3.31	1.01	0.62	0.40	490	35.03	10.05	0.20
Chicory	1.11	5.51	1.07	0.92	0.37	469	57.05	12.05	0.20
Dandelion	0.30	3.59	0.47	2.21	0.64	619	48.00	18.02	0.13
Red fescue	0.57	2.19	0.27	0.46	0.05	275	26.02	10.03	0.19
Meadow fescue	0.58	2.47	0.44	0.66	0.20	250	29.02	9.05	0.16
Trefoil	0.79	2.77	1.27	0.56	0.18	383	43.04	7.03	0.20
Timothy	0.44	2.74	0.36	0.70	0.17	408	25.03	7.06	0.15
Alsike	0.76	2.59	1.04	0.42	0.06	323	58.08	10.05	0.17
Perrenial ryegrass	0.59	2.38	0.35	0.51	0.19	252	21.06	8.05	0.15
Cocksfoot	0.59	2.78	0.36	0.31	0.17	200	45.05	10.00	0.14

P (phosphate), K (potassium), Mg (magnesium), Cl (chlorine), Na (sodium) expressed as percentage of dry matter.

Fe (iron), Mn (manganese), Cu (copper), Co (cobalt) expressed as p.p.m. (parts per million) of dry matter.

Table 6.2 Trace-element requirements of livestock (parts per million of dry matter)

	Cattle	Pregnant cattle	Sheep
Iron	30	30	30
Copper	10	10	10
Manganese	40	40	40
Cobalt	0.01	0.01	0.01
Molybdenum	Known to be detrimental to health		
Iodine	0.12	0.08	0.08
Zinc	50	50	50

mixtures are the root cause of their deficiency problems. The seed mixture we were recommended to use is detailed in table 6.3, and, apart from some slight differences of variety, is the same as we use today. This is how the experts explained their recommendations:

> Comparing the livestock requirement with the mineral content of the herbage you will note that all the components of this mixture provide sufficient iron and cobalt. Assuming that the cocksfoot provides one third of the bulk and the perennial ryegrass two thirds, then the Manganese content averages out at 30 p.p.m. The requirement is 40 p.p.m. As you will note this will be made up mainly by the chicory and partly by the burnett. The copper content of the cocksfoot and ryegrass averages out at 9 p.p.m. Again the chicory will make up the 10 p.p.m. requirement.
>
> The mixture itself is suited to the system although close cropping is needed to keep the cocksfoot under control. The cocksfoot which is very deep rooting will conserve the soil structure, and at the same time restore the organic content of the soil.
>
> The creeping red fescue is included as a bottom grass which will also bring the copper content up.

Table 6.3 Recommended seed mixture

Species/variety	Pounds per acre
Perennial ryegrass (late) S23	6.0
Perennial ryegrass (early) S24	6.0
Cocksfoot S37	2.0
Cocksfoot S26	1.0
Cocksfoot S143	1.0
White clover	0.5
Wild white clover	0.5
Timothy S352	2.0
Creeping red fescue	2.0
Burnett	0.5
Chicory	1.0
Sheep's parsley	0.5
TOTAL	23.0

Organic husbandry

I do not pretend that anything that we do is the only way to farm organically. For instance, I am very conscious of the fact that, while we do have a lot of stock of various sorts, we have no dairy cows and so do not appreciate the problems that they would present. But I believe that the basic principles of good husbandry apply to whatever type of farming is practised, and in this book we are dealing with just one farm.

7

The joint sheep enterprise

I decided very early on in this experiment that sheep were an essential ingredient of an organic farm. All the downland of Wiltshire was once a huge sheep walk and Wiltshire even has its own breed — the Wiltshire Horn — although it is something of a rarity now. The snag was that, while I like the effect of sheep, I think that as animals they are very stupid! Unless you are a good shepherd your flock will never flourish, and so I hit on the idea of starting a joint sheep enterprise. We needed sheep on the land and, being lucky enough to have a largish farm, we felt we could start a young man in farming without his having to find a vast amount of capital.

In essence, the enterprise works like this: a young man, the shepherd, supplies the sheep, the equipment and the labour, and we provide keep for them throughout the year. As the original advertisement stated, 'The enterprise is designed for a young man with some capital who wants to gain a foothold in farming, but who has insufficient capital to start on his own.' The shepherd pays a headage payment — currently 30p a week per ewe, and 12p a week for lambs from a date six weeks after lambing starts until 1 October, when they are deemed to be adult and are charged at the full rate — and uses land that is assigned to him by the farm manager. However, like all good stockmen, the shepherd always seems to know where the best bite is and the sheep find their way there before we know anything about it! The old story of a carter who was being interviewed by a farmer for a job comes to mind. 'Would you steal for your horses, my man?' 'Oh no, sir.' 'Then you are no good to me.'

Sheep being moved down the farm drove

What is meant to happen is that the sheep have a block of grass from April till the end of July where they lamb and get strong. For the rest of the year they go round the farm, clearing up behind the beef cattle and horses, grazing the new leys in the autumn and any roots we have left after shooting. The shepherd has to provide his own fencing, as we only fence against cattle, and he is very successful at containing them with three strands of electric fence wire. In addition to the headage payment we do get the benefit of the sheep (the 'golden hoof', as the old saying has it) on the land — sheep can be run on pastures in the wettest weather without ruining it, whereas cattle tend to poach it very badly and it takes a long time, without the help of bag nitrogen, to recover.

At the present time the shepherd, John Dewey, has about 750

98

lambing ewes on the farm, and lambing usually starts at the beginning of April. He has been a bit unlucky with his timing since he has been with us. Once he decided to lamb early (beginning of March) and was hit by a blizzard. The following year, 1981, he decided to lamb late (mid April), and was hit by another blizzard, the one that knocked down so many power lines on the night of 25−6 April. I have a photograph taken on 1 May that shows cowslips and a snowdrift, proving how bad and how late it was. Another slight hiccup occurred one year when the shepherd put out a 'teaser' ram, guaranteed impotent, and had some 200 ewes made pregnant by it! But all is not disaster — the sheep do very well on the fertilizer-free grass and the lambs all turn out well. One thing we do notice with the stock on our organic fields is that, although there does not appear to be a lot of growth, the stock always seem to be content. Presumably the dry matter is very much higher, so they do not need to eat so much.

As already noted, one of the most difficult things to accept on this system is the lateness of spring growth. Without bag nitrogen to speed the process, we have to wait for the soil to warm up so that the micro-organisms can do their work. One virtue the organic farmer has to learn is patience; but this is rewarded when, for example, one sees how well the stock do on the leys.

The sheep that the shepherd runs are Mules, which as the name implies, are a cross. The original cross was a Blue-faced Leicester ram on a Scottish Blackface ewe, but a Blue-faced Leicester ram with a Swaledale ewe is also classed as a Mule, just to complicate matters. The main advantages of this cross are said to be that it provides the best mothers — keeping to a minimum losses through mismothering — and that the ewes have plenty of milk to take twins through to the fat lamb stage. When sheep were kept on the farm before, the breed that was then favoured was the Clun, but now the Mule is the most popular in this part of the country. The shepherd uses Suffolk and Dorset rams and the lambs look really well by September time.

The agreement with the shepherd runs from 1 October each year. Either party can give notice of termination by 1 July in any year to become effective on the following 1 October. In this way a young man who is hoping one day to have his own farm or to move

Above: A flock of Mule sheep
Left: Portrait of a Mule

elsewhere is only tied for one year at a time, and if the contract is to be terminated there is plenty of time between 1 July and 1 October for him to arrange to sell his stock if he wants to. Similarly, the farmer has time to advertise for another shepherd. John Dewey started with some 300 ewes in 1974 and is still with us in 1986, which maybe speaks for itself.

During the winter, after the sheep have cleared up all the grass and roots (kale or swedes), they usually spend a month or so on the edge of the Larkhill ranges and the shepherd is given a bale of hay for each sheep. This, together with a few pounds of hard food before lambing, is enough to take them over until the grass comes, but, as I

Above: Waiting to be shorn

Right: John Dewey shearing

mentioned before, they seem to be chasing every blade of grass in May.

The benefit of this type of system is that a young man with insufficient capital to start a farm on his own can run his own sheep on someone else's land. Apart from the headage payment to the farmer and the other obvious costs, the balance is profit for him. Obviously the level of the headage payment is critical and the farmer must set it at a figure to give the young man plenty of encouragement to work hard and make as much money as he can. To all intents and purposes, he is his own boss as far as the sheep are concerned — we control where he goes and we do insist that he keeps his area of the farm tidy. With a young man this is probably the most difficult

Newly shorn!

thing to instil, but, as we have so many visitors and we keep the rest
of the farm tidy, we do make sure that he does too. The benefit from
our point of view is that we have the sheep on the land without the
worry. We used to run a flock of sheep ourselves, and it always

seemed to me that, when they weren't dying, which they seemed to do with depressing frequency, their demands always coincided with the peak times on the arable side. Lambing came at spring drilling time, shearing and dipping at hay-making, and drenching and sorting for the sheep fairs in harvest. While on our chalk hills you can see to the inch where sheep are penned, it is nice to have someone else doing the day-to-day job of looking after them, and in my book that means the shepherd deserves the profit.

One hears so much about would-be employees writing dozens of letters for jobs only to receive no reply from employers that it may be interesting to note our experience. When we advertised originally for a young man (you were allowed to stipulate the sex of the person you wanted in those days!) to take part in this enterprise, we had about fourteen replies requesting details, and these were duly sent, but over half didn't bother to write again, either to say they were not interested or to explain that they had found another job. Quite recently we were contemplating a joint enterprise for free range eggs, run on somewhat similar lines to the sheep enterprise. Eleven people replied to our advertisement in the farming press, and all were sent full details. Of these eleven, only one bothered to reply, so I think the fault is not all on one side. Incidentally, the free-range-egg idea fell through as the only serious applicant felt he would prefer to work nearer home.

As it may be of interest to others thinking of embarking on a similar scheme, the actual proposal that was sent to all the applicants for participation in the joint sheep enterprise is printed in Appendix E. Many details have changed since 1972, when the original advertisement was placed — in particular the headage payments and the shift away from fertilizer — but the document gives a fairly accurate idea of how we set the scheme up and how it still operates.

8

Visitors and estate work

To complete the picture of what we are trying to do at Rushall, I now turn to the subject of estate work and how we cope with visitors. Whether one likes it or not, the time when you could order people off your land just because they were there is rapidly disappearing. The huge urban population are coming to regard the countryside as their playground — for themselves, their children and their dogs. While they would probably be the first to shout if we started wandering over their lawns and turned our dogs into their gardens for their daily run, they seem to regard farmland as theirs, and cannot understand such things as damage to crops and disturbance of livestock and wildlife. Farmers could and should do more to educate these people so that a mutual respect is established for the benefit of all concerned.

At Rushall we entertain many people throughout the year from all walks of life and with many different interests. Some are private guests, some members of clubs, some customers and some simply interested people. All, we hope, gain something from seeing what we are trying to achieve, and we, for our part, increasingly get the message that people wish to see more of the countryside than they do from just driving through it in their cars, and that they genuinely want to understand more about it.

Our most distinguished visitor has been, without doubt, HRH the Prince of Wales, who came to look round the farm and mill on 5 November 1984. He showed a real interest in what we are trying to do and spent about two hours with us before moving on to the Elm

The home of Rushall flour

Farm Research Trust, near Newbury, to look at their work and to meet several members of the organic movement. His concern for the future of the environment is well known, as can be seen from the following two extracts from speeches he has made. Addressing the annual conference of the Conservation Society at Cheltenham on 17 April 1982 he said,

> We cannot sensibly talk about conservation and at the same time promote a way of life that consists in the ever more accelerated stripping of the earth of the very materials needed to sustain life.
>
> Those of us who believe that we have an over-riding responsibility to leave the planet Earth in a condition at least

no less capable of supporting life than that in which we found it, have no reasons for believing that mankind is discharging that responsibility to future generations.

There is an accumulating body of evidence which reveals that the predominant forms of industry and agriculture in the industrialised world are non-sustainable.

We would be deluding ourselves if we believed that a technology based on an abundant supply of cheap fossil fuel can long continue and it is much more unrealistic to expect it to spread throughout the Third World.

Neither the resource base, nor the environment could stand it.

We need to find constructive ways of changing the direction of science and technology into more harmonious channels by encouraging much greater understanding of natural — and sustainable — biological systems.

And in a message to the Organic Growers and Farmers Conference at Cirencester on 8 January 1983 he remarked,

For some years now modern farming has made tremendous demands on the finite resources of energy on earth. Maximum production has been the slogan to which we have all adhered. In the last few years there has been increasing realisation that modern farming methods are not only very wasteful but probably also unnecessary.

A rather amusing incident occurred on the morning of his visit, before he actually arrived. The police were making a routine search of the premises when my wife had occasion to call our son's dog. This black labrador was with Stephen in London before he moved to Paris, but because of import regulations he felt he couldn't take the dog, who was left with us. He is not always the most obedient of animals and often has to be called several times and in rather a loud voice. His given name is Charles Le Chien, but, as can be imagined, this is almost invariably reduced to 'Charlie'. So when, on the morning of the royal visit, my wife started calling 'Charlie, Charlie', one of the detectives came up to her and said quietly, 'Madam, today I think we will call him Rex.'

Our other visitors (friends, customers and business contacts apart) can be divided into three categories, although these obviously overlap: those with sporting interests, naturalists, and those interested in the farm and mill. Some 3000 visited the farm for one reason or another during 1985, and I hope that at least some of them have gone away with the feeling that not all farmers are vandals, nor are all organic farmers mad. So far we have had visitors from some thirty-four countries, as far afield as Argentina, China, New Zealand and the United States.

The sports that attract people to the estate are the usual country field sports of hunting, game shooting and fishing, plus horse trials and clay pigeon shooting. There is a thriving clay-pigeon club that uses an old chalk pit on the estate. This club has been operating for several years and at present has about fifty members. They shoot on alternate Sunday mornings and occasionally on weekdays during the summer. It is an expensive hobby but one that has its devotees and they certainly are expert. The game shooting is largely pheasant shooting at the moment with a few ducks and the occasional partridge. One of my fondest hopes is to demonstrate that the wild English partridge can still flourish on a farm that is free of all chemicals. To date we have nothing to show for our efforts — we have a sprinkling of these lovely birds, but they just don't increase. We have had three really bad springs lately, with the summer of 1985 being particularly hostile to young partridges, so we are still hoping that with a few good springs followed by reasonable summers we may yet see a sufficient surplus of partridges to be able to have a partridge day again. Meanwhile, we rear a few pheasants to augment the wild stock to enable us to have about eight to ten days shooting, some of which we sell to help defray the costs. Even on a small estate such as ours we find it necessary to have someone around all the time to help the manager keep an eye on things, but we cannot afford to keep someone just as an estate policeman. A few years ago I had the idea of having a man as estate policeman, woodsman, river-keeper and gamekeeper all rolled into one, and advertised for a 'countryman'. We had about 200 replies, which is infinitely worse than getting two! The applicants varied from gentlemen of dubious background to gentlemen of dubious worth! Eventually we settled on one man who was with us for about two years before being pirated by the Game Conservancy, of which I am a member, to run its experimental beats on Salisbury Plain. While this man was with us it became apparent that the job was too comprehensive, so now we are back with a gamekeeper who keeps an eye on the fishing and the estate in general. With eight guns a day shooting you can reckon we entertain about eighty to a hundred people in the course of the season, as well as providing a little casual work for any extra beaters and pickers-up we may need.

It is interesting to compare the numbers of people who enjoy the estate by way of hunting and shooting. With shooting we have a maximum of a hundred; with hunting, if the hounds visit us four times a year and there are sixty followers each time, a total of 240. One can argue endlessly about the relative merits and attractions of the two sports, but it is essential for them to coexist, neither being so jealous of its own interests as to exclude the other. They are both part of the country scene, and it would be a sad day if either were ever banned.

The expanding sport of horse trials is also one of our interests. From 1970 until 1985 we organized the Rushall Horse Trials on our land, and we have calculated that over those years we started about 2700 competitors on our cross-country courses, with 79.35 per cent of starters completing the course. My wife has written a lively account of the 1979 trials, and to give some of the flavour of the occasion, with all its problems, frustrations and rewards, I quote here part of her account of the last day:

Looking back it is impossible to recall the day in detail. It is a blurred confusion of isolated incidents. Two loose dogs and Barry's announcement that they could be collected from the secretaries for a fine of £20. Sue Hatherly turns up to claim one, with a rather worried grin — was the fine meant seriously? Barry said no, but he would collect if she won a prize. No one seemed to think the funny little friendly white terrier with a long back and short legs was worth £20, so we tied him up in the office at the drier. A telephone call from home to say the caterers were on their way after all. What on earth could I say

111

to them if they arrived in time? Another harassed call from the dressage steward, Joyce Hayward — they were short of Intermediate Dressage sheets. Barry rushes Marilyn [the secretary] back home to the typewriter once again. They return remarkably quickly, only to find that Stan, at that time controlling the parking at the drier was needed urgently to repair a blocked pipe, so all 6½ stone of Marilyn takes over as car park attendant.

Barry calls in and asks if I'd like a quick trip on the cross country, Karen seems fine and Marilyn is back from controlling the traffic, so I guiltily slip away. So far the course is riding well. A few quick words with brother Thomas, standing on France bridge and politely lifting his cap as each competitor shoots by 'Careful, bridge is slippery', and invariably receiving a hurried 'Thank you' in return. An offer of some rather revolting coffee from son Nigel's double decker bus, where he is feeding the baby and his wife is peeling sprouts and roasting pheasant — surely this, at least, must be unique for a horse trial? A hasty visit to Morris Nicol in the thatched pavilion, who looks after the doctors and vets. He's bored, thank heaven. Back to work, and no one has yet claimed the terrier. The announcer feels he needs a walk. . . . Someone wishes to see me I'm told. A smartly dressed young country gentleman approaches — 'Special Branch' he whispers in my ear, and I feel terribly important. By now there is a colossal din in the drier. The bar is going full swing and everyone is shouting. The scoreboard writers are finding it very difficult to hear Gina's voice from the score tent. It's past mid-day and it still isn't raining. For some extraordinary reason I am suddenly presented with two pages of show-jumping scores, expressly intended for me I am told. John Wallis, one of the commentators, is at hand and says he needs them in the commentary box, so relieves me of them with the explicit instruction to check with the scorers first. He overlooks this minor detail, while the scorers are madly ringing and dashing in all directions to locate them. And for once the show-jumping officials have omitted to use their carbons! I notice the cherry brandy tucked in a corner. Whom can we spare to dispense the customary cheer to the jump judges? . . .

No rain until 3.30, when I am once again on the course watching the Advanced Class. Not so guilty this time as Karen and Jan prefer to watch the National, and Marilyn isn't interested in horses. Everyone is enthusiastic and full of thanks — goodness knows how some of them managed their times. It's nearly over and we still have the terrier. Everything seems to be running to time, and the Red Cross ambulances haven't moved. Barry is at last relaxed and offering liquid sustenance to all and sundry. Nigel says he will take the terrier home as he is worried about him. It may be a solution but they already have a number of sheep, a dozen assorted hens, one pony, one cow, two calves, one pig, two cats, four dogs and a baby. The terrier doesn't appear worried and is taken for his third walk.

Prize-giving. Thanks from Tony Wootton, always very graciously given, and an unexpected tribute from Mr Boreham. Can't think why the Midland Bank should thank us. Sue Hatherly comes second and gives Barry a sideways glance.

In no time at all it's all over and place is almost empty, apart from the terrier. I suddenly remember the police, and dash up the ladder where they were chatting to Barry. 'They went off in a hurry,' someone says. 'They were afraid you were going to ask them to take the terrier.' I was.

Why do we do it? I don't know. In retrospect it's fun. We make a lot of friends and we laugh a lot. We are also frequently exasperated and always worried.

Other horsy events include a hunter trials for the local riding club and a sponsored ride for those who are not competitive. All in all we must have several hundred horses on our land every year.

Fishing for trout is a very different sport. One definition of the word 'sport' which I once read said that it was a pastime containing an element of danger. Whether the possibility of having a fly stuck in your ear is sufficiently dangerous for fishing to be called a sport is open to question, but of the three field sports that I have been privileged to enjoy I think fishing is probably my favourite. To start with, it is best conducted by yourself with no chattering companions, who can ruin a day's hunting or shooting. It takes you to some of the more beautiful parts of the country and water is always fascinating even if no fish oblige. I have been fortunate enough to fish for several years in Norway, where the scenery is truly magnificent and so are the fish! I am told I am the only living Englishman to have caught there two salmon weighing together over 100 pounds (they were 55 and 47 pounds respectively.) At Rushall we have the two branches of the River Avon, prosaically called the Western branch, which rises at Cannings, and the Eastern branch, which rises at Wootton Rivers, near Pewsey. Both are very narrow and the trout need a careful approach if you are to take the limit of six. We have a small syndicate of friends to fish the rivers, and the only significant rule is that they treat the estate as if it were their own.

Naturalists seem to be a rapidly expanding brotherhood. Each year we have more and more requests for someone to come onto the farm to study the birds, the insects, the flowers or the butterflies. At one time I tended to try and discourage these people for fear of disturbance to the game or damage to the crops. Experience has shown me that this worry was largely unfounded: nearly all these visitors are very careful to cause no disturbance and the results of their labours are very informative. Not being a naturalist myself, I have been surprised at the number of wild flowers, butterflies and moths they have discovered that I had not noticed. My eyes were really opened one day when the wife of a friend rang up to tell me that I was not to graze a particular piece of downland for the next month. Not surprisingly, I demanded to know why, only to be told that there was a burnt-tipped orchid growing there and it wasn't to be disturbed! When I had simmered down a bit I persuaded her to come over and show me, which she did, and my interest has been growing ever since. We can now recognize three types of orchids on one of our farm roads, and they seem to be spreading, much to the delight of our summer visitors. In my callow youth all wild flowers were regarded as weeds and, if we could find the time, all the tracks were sprayed to kill them. But as one gets older the number of wild flowers that fall into the category of 'weeds' becomes progressively less. However, I would need to live a long time before I looked on docks, nettles and thistles as anything but weeds. We also have a

114

A common spotted orchid growing alongside the farm road

withy bed that grows a lot of comfrey, which in turn supports a colony of scarlet tiger moths. To be honest, although I have been down to see them on several occasions it has always been too cold or too wet or too something for them to be out. I am reminded of a ghillie who once said to me of salmon fishing, 'When everything's right, something's wrong'!

Visitors to the farm and mill account for most of the 3000 visitors we had in 1985. The mill manager, David Fuller, and his wife Sue have given innumerable baking demonstrations to groups of young wives, members of the Women's Institute, school parties and many others. Those with more interest in the farm are taken in hand by the farm manager, Richard Fry. It has been very noticeable that the number of *bona fide* farmers who come to have a look at what we are doing has increased a lot over the last few years. As they find the present high-input system becoming less rewarding economically, they are looking for alternatives, and many express the view that they would like to do something similar if they thought they could afford it. (This point is discussed in chapter 9.)

We have a lot of school visits and the children always seem to enjoy themselves. David Fuller shows them how to bake bread and before they leave they are given a roll that they have seen made. This lasts about as far as the door! What does surprise us is that so few of the country children seem to know the link between growing wheat and eating bread. Presumably most of their mothers shop in supermarkets where bread just appears on a shelf and they never question how it is produced. We usually receive some fascinating letters from these children when they return to school — one of them is reproduced here.

One way and another, then, we entertain a lot of people during the course of a year. It is rather time-consuming, and in the summer months, when we are very busy, there are moments when we wonder whether it is worth it. The view which usually prevails is that, if our visitors are genuinely interested, we should do our best to accommodate them, as I believe example is better than precept. They can see and judge for themselves whether the system is viable and sufficiently professional for them (if they are farmers) to contemplate having a go themselves. We always try to fit the children in because of their almost total lack of knowledge about the simplest farming operation, and before very long they will be the parents of yet another generation. In addition to these visitors we have a growing number of people, mostly students and reporters, who ring

116

All Cannings School,
All Cannings,
Devizes,
Wiltshire,
October 23rd

Dear Mr Fuller,

Thank you very much for the visit to the mill. I enjoyed it very much and so did the rest off the school. I thought it was very interesting how you made the animals out of the dough and also how the machine made the dough into rolls. I also liked the taste of the rolls after they had come from the Oven. I also thought that splitting up the corn is a good idea so they don't get mixed up with each other because each bit of grain is different. Anyway thanks you for the visit it was a lovely afternoon out. Thank you very much.

Yours Sincerely
Amanda Hambleton.

up and expect to be told, in five minutes flat, how to farm organically. Perhaps I shall now be able to tell them to go and buy this book!

Away from the visitors, a lot of work has to be done to maintain an estate. Some aspects of this, such as building maintenance, are pretty well the same wherever you go, but in an estate's woodlands, as in its fields, you can see the personal touch of the owner. It is said (and I accept it as a compliment even if it is not intended as such!) that I go 'tree mad' every spring. There is always so much to do in the woods at this time of year — felling, planting, pruning, thinning: the list seems endless — before the leaf starts breaking and you can still see what you are doing. Planting trees is the easy bit; it is their maintenance for the next twenty years that entails so much work. But it is very rewarding: nothing is more pleasurable in spring than to walk through a well-tended wood, with trees of all ages and varieties, just as they are coming into leaf. I remember the man who produced the Benedicite window in Rushall Church telling me that the company he worked for had a bank of stained glass containing 3000 shades of green. The doubter has only to start looking at the countryside in spring to realize that such variety is no figment of the imagination.

Basically, what we are doing with our woodlands is to clear all the old woods which contain dead and dying trees and replant them with a mixture of soft- and hardwoods. The conifers are useful as a source of cover until the hardwoods are strong enough to stand on their own. They are then progressively removed for fencing material, leaving a stand of hardwood — oak on the vale land and beech on the chalk — to mature. Any self-sown sapling found, with the exception of sycamore, is always protected with a plastic rabbit guard and encouraged to develop. I only recently discovered that the sycamore is not an indigenous tree but an import of comparatively recent times. It is very prolific and grows from seed like weeds. Care should be taken not to leave too many sycamores as they do tend to

A stand of poplars, planted in 1966

kill other trees in their vicinity. But a sycamore is better than no tree at all.

Dutch elm disease removed all the elms on the estate and caused the tree programme to be thrown out of gear. It took us about two years to fell and remove all the elms on the farm and little else was done to the woods during that time. We used as much elm timber as we could for posts (not a good idea, as they don't last long) and as waney edge (see Glossary) for cladding on buildings. This is, or was, a traditional way of building in Wiltshire and with old bricks makes a very pleasing building. To replace the elms we have used beech, oak, horse chestnut and Norway maple with a few lime and larch, but it will be a long time before they stand as tall as the elms they have replaced. It will be very interesting to see whether the young elms which have sprung up since the old ones were felled will ever reach maturity. There are many now of about 12—15 feet high, and, while a few fall victim to Dutch elm disease each summer, a surprising number seem to escape and keep on growing. I am told that a slightly less virulent form of the disease struck the South of England some sixty to eighty years ago, and obviously many trees must have survived and many young ones grew away from the roots of the old. Let us hope that the remaining trees will somehow develop immunity and that we shall see the rooks nesting in the elms again before too long.

The cottage situation is one that has caused us some concern over the years. We need a certain number for the men on the farm and we always try to house retired men and their widows. The rest are let, mostly to local people, with particular preference to young newly-weds who have no hope of a council house, as they obviously cannot have many points in the priority system employed by the local district council. Some years ago, we found we were being forced into court actions to repossess our houses. This process is always rather upsetting, so we devised a scheme called the Farm Cottage Scheme and persuaded the local council to adopt it. The objectives were threefold.

Stocktree cottages, Rushall

1 To encourage the letting of surplus farm (or other) cottages.
2 To take people off the local council's housing list.
3 To enable the owner to repossess his house without recourse to an eviction order.

Under this scheme, the owner has to undertake to let his house to someone from the council waiting list (he can choose a tenant from their list) for a period of at least one year. After that, should he require his house, the council undertakes to house the tenant within a period of two months. This avoids the need for eviction orders, but it does in effect mean you have to let your house for a minimum of fourteen months, although in practice the length of tenancy is considerably longer.

The scheme has worked very well in the local area, and has been copied or adapted by several other councils in the country.

To
our grandchildren
KIAN, JETHRO and RACHEL
and to all future farmers

Contents

Foreword

Lord Penney, OM, FRS

I first met Barry Wookey a few years ago at a dinner where he was the guest speaker. He spoke about organic farming and his experiences at Rushall Farm. Now, at his invitation, it is my pleasure to introduce his book on the same subject and to offer my own views on why organic farming is so important.

Farming is man's most necessary industry and, like all industries today, it must consider whether its normal operations are damaging the environment or are causing potential risks. In the past, people had to endure whatever came and only when the situation had become intolerable was anything done. The pollution of the Thames and the Rhine, the London pea-souper, the smogs in Los Angeles and other big cities are examples of hazards against which successful action has been taken. Today, and for the past thirty years, atomic energy has been under close public scrutiny in regard to pollution, health and safety. The effects of acid rain caused by sulphurous emissions from fossil-fuel power stations are now being intensively studied and some actions have already been taken or planned. There are other cases where questions are being asked. One of those relates to farming, and the possible risks resulting from the use of chemicals in farming. Some farmers, recognizing those risks, have converted to 'organic' farming, which uses no chemical fertilizers, no pesticides, fungicides or herbicides, does not inject animals with hormones and has no place for battery hens.

In most countries, a key indicator of food production is cereal production. Over the last decade, world production of cereals has increased a little faster than world population. In two large regions, Africa and the Soviet Union, cereal production has not increased and may have fallen. Elsewhere, there has been a general increase, leading in some of the exporting countries to economically destabilizing surpluses.

Cereal production in the European Economic Community over the past decade has increased much faster than the world average and food mountains have appeared. Cereal production in the United Kingdom on three-year averages increased by 60 per cent during the eight-year period 1974–6 to 1982–4. (Wheat production over the same period actually increased by nearly 150 per cent!) The corresponding increases in cereal production in France (easily the largest EEC producer), West Germany and Italy were 40, 17 and 10 per cent, respectively. There were three main reasons for the increase in production. First, new high-yielding, short-stature cultivars were introduced; second, there was increased use of chemicals on farms; and, third, the EEC's Common Agricultural Policy offered financial incentives for cereal production.

In the United Kingdom, the greatly increased use of home-grown cereals for animal feed is a consequence of the heavy application of nitrogen fertilizer to the new cultivars, producing larger quantities of low-quality grains. Both in the United Kingdom and elsewhere in the EEC, there is concern now not only about the food mountains but also about agricultural (and other) chemicals getting into the water supply through seepage and emissions.

Past experiences in several countries emphasize the need for caution in the use of agricultural chemicals. The use of the organo-phosphorus insecticide DDT was abandoned because it was persistent and was accumulating in the bodies of some types of wildlife. Traces were being widely found in man. The introduction of new pesticides after DDT at first was proved effective, but resistant strains appeared and doses had to be increased by two to three times, leading to the production of new types of pesticide. The struggle continues and becomes ever more sophisticated. Nature always fights back.

Some of the chemicals used on farms get into rivers, lakes and underground waters. Lakes have been damaged by an over-supply

9

The economics of organic
cereal farming

I am indebted in this chapter to the work of Helen Browning, who, with ADAS support (see chapter 11), produced figures for 1983 and 1984 showing how Rushall and neighbouring farms that are being farmed conventionally compare in terms of the economics of producing wheat. Figures being what they are, in the past I have hesitated to use them for fear of being thought over-optimistic or too subjective; but now that the ADAS figures are available it is hoped they will be accepted as completely objective, and, while they may surprise some people, I trust they will be found helpful. After all, farmers are businessmen and they have to make their businesses pay, so, however much they might like to farm organically, if they would go bankrupt in the process they can hardly consider it.

Before going into the details of the ADAS comparison we should, perhaps, discuss the question of organic premiums. It is often said to me, 'It is all very well for you because you sell your wheat and flour at a premium price. If everyone did it there would be no premium, and then you couldn't exist.' On the face of it this seems absolutely fair comment. We do enjoy a premium of approximately 100 per cent for our wheat and flour over high-yielding feeding varieties which are grown with the full range of fertilizers and chemical sprays. Without that premium, I for one, could not have survived. But there are two other important points I should like to make. The first is that for a long time the agro-economists, if that is what they are called, have been exhorting farmers not just to go on producing

A typical crop of winter wheat — spot the weed!

and hope that someone will buy whatever and however much they produce, but first to find out what the markets want and then produce for them. That is exactly what we at Rushall have done, and *the market pays a premium because our produce is wanted.*

The second and more important point is this: conventional farmers, growing vast quantities of low-grade wheat that no one seems to want, enjoy a remarkable 'premium' from the taxpayer. I have calculated that for an average (3.5 tonnes per acre) crop of winter wheat, sold into intervention and stored for six months before being sold on the world market with the necessary export restitution, attracts no less than £184 an acre (based on prices prevailing for the 1985 harvest) — in effect a subsidy paid by the taxpayer to the farmer. The calculation is presented in table 9.1.

So all comparisons between the two systems start with this basic difference: the conventional farmer is supported by the taxpayer by way of intervention, and the organic grower relies entirely on the market with no help from the Government. This should be borne in mind when looking at the figures, because, if conventional 'chemical' farmers had to rely on open market prices for their wheat, few, if any, would be viable. This leads on to another important point: if many farmers were to adopt an organic approach, the premium would eventually reduce to a level at which the system would cease to be viable. At what level of production this would occur is debatable. Some say the market could absorb up to 20 per cent organic produce at present price levels, although I would think 10 per cent would be more realistic; but organic producers should

Table 9.1 Cost to taxpayer of feed wheat held in intervention store for six months before sale on world market, 1985—6

	£
Price per tonne of wheat delivered to intervention, store, August 1985 (price paid by HM Government, which then owns the wheat)	111.01
Storage for six months (at £1.59 per tonne per month), paid by HM Government	9.54
Total cost per tonne of wheat held in intervention stores for six months	120.55
Less world market price per tonne of feed wheat in February 1986 (109.66 ecus; £1 = 1.61641 ecus,[a] February 1986)	67.84
Deficit per tonne to be made good from EEC funds	52.71
Total deficit per acre, at 3.5 tonnes per acre	184.49

[a] ecu = European currency unit

accept that over-production will bring with it problems which do not exist at the moment, and that in time it may well be necessary for the Government to support organic farming in the same way as it supports conventional farming now.

The ADAS study used normal feed and milling prices for our organically produced wheat 'in order to assess the two systems on an equal basis'. The calculation does not, therefore, reflect the actual returns from our wheat-growing fields, as none of our wheat was sold at less than £200 a tonne, whether as wheat or as flour. The 1986 prices for our wheat and wholemeal flour are listed in table 9.2.

Without going into all the details, the costs for cultivating and chemical inputs have been calculated at the average pertaining at the time. For instance, fertilizer prices have been calculated on the basis of 36.6p per kilogram for nitrogen, 32p per kilogram for phosphate and 16p per kilogram for potassium, plus a 10 per cent compounding charge; adjustments have been made where sprays were tank-mixed. We must, I think, assume that ADAS has done its part quite impartially, which makes the final figures, presented in table 9.3, all the more valuable.

The conclusions drawn from these figures by ADAS were as follows.

Table 9.2 1986 prices (from 1 October) for Rushall wheat and wholemeal flour (ex-mill)

	Price (£)
Flour	
25 kg bag	8.00
12.5 kg bag	4.50
1.5 kg bag	0.65
0.5 tonne lots (pertonne)	275.00
1 tonne lots (per tonne)	250.00
Wheat	
1 tonne lots, in bulk (per tonne)	200.00
1 tonne lots, in customer's bags (per tonne)	230.00
1 tonne lots, in bags supplied (per tonne)	240.00

Table 9.3 Average margins for twelve wheat crops (six organic, six conventional) in 1983 and 1984.

	Conventional	Organic
	£ per hectare	*£ per hectare*
Seed	40.60	48.80[a]
Fertilizer	118.28	—
Sprays	96.43	—
Cultivation costs	65.38	75.72[b]
Harvesting costs	58.00	58.00[c]
TOTAL	378.69	182.52
	Tonnes per hectare	*Tonnes per hectare*
Average yield	7.4	4.4
	£ per hectare	*£ per hectare*
Output at standard prices[d]	801.60	506.00
Output at organic premium price (£160 per tonne)[e]	n.a.	704.00
Output at minimum actual organic premium price (£200 per tonne)[f]	n.a.	880.00
Margin at standard prices[d]	422.91	323.48
Margin at organic premium price	n.a.	521.50
Margin at minimum actual organic premium price[f]	n.a.	697.48

[a] The difference in cost reflects the fact that we use varieties that are no longer in common use and so cost more to buy. We normally save our own seed, but this has not been allowed for here.

[b] Several of our neigbours were using direct drilling techniques, so reducing the number of cultivations necessary. Ours included extra passes to prepare a weed-free seed-bed

[c] The equal costs seem a little hard on the organic farmer, whose yields are so much less!

[d] Where a variety is suitable for milling it is assumed to have been sold as such.

[e] The price of £160 per tonne for organically grown wheat was the premium price being paid by a merchant dealing in organically grown grain.

[f] Not included in the ADAS report; included here for ease of reference.

Organic cereal farming

1   In 1983 and 1984 the average margins from the organic wheat enterprise at Manor Farm [Rushall] were approximately £100 per hectare higher than those achieved by the monitored conventional crops, due to:

(a)   The premium available for organically grown wheat.
(b)   An average saving of £216 per hectare by the organic farm on inputs, i.e. sprays and fertilizers.

At conventional prices, margins on Manor Farm would have been approximately £100 per hectare lower than those achieved by the conventional fields.

2   Seed and cultivation costs were slightly higher on the organic farm; scarcity of some varieties, such as Copain, and the extra cultivations needed to prepare a weed-free seed-bed on the organic farm accounted for these extra costs.

To these I would add the following observations.

1   The 1983 and 1984 harvests were good. 1985 was terrible and 1986 (after the disastrous February) was nearly as bad, so all these figures should be read with caution. In other words, don't expect margins of £697.48 per hectare every year from every crop of organic wheat!

2   The organic farmer cannot grow continuous wheat. The figures reproduced above compare fields that happen to be in wheat in any one year — the whole-farm figures are those that really count.

3   The premium we currently enjoy may well have to be replaced by direct Government support if and when there is sufficient organically grown produce to affect the market.

4   We have been able to undertake this experiment because we had sufficient land to carry it. I think it would be extremely difficult, if not impossible, for a small arable farm, with high rent and a high overdraft, to convert to an organic

Growing barley

128

system without financial help from the Government. (See chapter 12.)

I remember that when I took over Rushall in 1964 I asked ADAS to advise on the beef enterprise that my father had been running; I knew nothing about it and wanted to know whether to continue with it. ADAS sent along its adviser and he went into all the figures relating to the ley system then in use, all the bought-in feed, and so on. After about a month he came up with the answer: every acre of barley was producing a profit of £30 an acre and every acre of grass was producing a loss (via the beef cattle) of £2 an acre. This shook me somewhat until I enquired how much credit was given to the ley for the cereal enterprise, as we used the ley to build up the fertility to grow corn, much as we do now. The adviser replied that ADAS didn't work like that: it took each field in isolation for any one year, and the resultant figures only reflected what had happened in that year. This seemed to me so blinkered that I immediately decided to expand the beef enterprise and we have never regretted it. This decision was probably helped along by a piece of tongue-in-cheek advice given me by my father when I started farming. 'Whenever,' he said, 'the experts tell you to do something, do the opposite.' I suppose my mind was already, in 1964, inclining towards sustainable systems of farming rather than the completely chemically dependent systems being advocated by ADAS. When this adviser told me, in effect, that rotational farming didn't figure in the Service's thinking I knew instinctively that it must be wrong. The whole-farm figures, including those for livestock, are what matter, and it is to the ADAS assessment of these for 1983—4 that we now turn.

Because of the joint sheep enterprise (see chapter 7), the ADAS calculations for sheep at Rushall had to be on the assumption that all the lambs were finished on the farm and the profit credited to the farm. For beef cattle, all of which are sold as stores at present, the actual figures were used. Figures for the horse enterprise, which was considered 'very specific' to Rushall, were not included.

Wheat in flower

Organic cereal farming

The overall figures should be considered somewhat provisional owing to the fact that at the time of assessment a number of fields were still going through the conversion process, when yields are very much lower than can be expected once the process is complete. Conversion is still continuing, in the sense that a number of fields have yet to achieve their optimum fertility. In ten years' time, if we are still in business, conversion will be complete and the figures for the various enterprises and so for the whole operation will be more reliable as a guide to the economics of an established organic farm.

Wheat cut with a binder for thatch with the combines cutting at the end of the field

The ADAS paper on the financial aspects of our organic farming operation concludes,

The whole farm situation — a summary

At the present stage of conversion, it is difficult to calculate an accurate GM [gross margin] for the whole farm as an organic enterprise. However, if the trend towards a 1:1 ratio of cereals: grass is maintained, and wheat remains the predominant cereal then it is possible to draw up a comparison for the whole farm situation.

Data from Exeter University would suggest a gross margin per hectare from beef and other cattle (excl. beef cows) of £341 per hectare (1983—4), and sheep £347 per hectare in conventional systems.

Farming on a 50 : 50 basis:

	Conventional (£ per hectare)	(£ per hectare, incl. premium)	(£ per hectare, excl. premium)
Wheat	423	521	323
Livestock (cattle and sheep)	345	233	233
Average farm margin per hectare	384	377	278

This analysis suggests that with the assured premium for organic wheat and at the present level of productivity from grass and forage crops, the returns at Rushall are comparable with average returns from conventional farms with similar enterprises.

There are many protagonists of organic farming who will say that its general adoption would solve all farming's present troubles. Others will tell you that yields are as good as, or better, than those obtained from conventional systems. My own view, having been trying to farm organically for the last fifteen years, is that we can only hope to produce about half the output of a 'chemical' farm. When talking to groups of farmers I often amuse myself by saying that they all produce the average 4 tonnes per acre of wheat which we hear so much about. Some of the best on the best land may achieve this (I am speaking of averages, not just the best field in any given year), but, judging by the very few who are prepared to look me in the eye when I say this, I suspect that many do not. Our aim is to average 2 tonnes per acre, but we have yet to achieve that. In 1984 it was 1.8 tonnes per acre and in the poor harvest of 1985 it was down to 1.4 (the ADAS result for 1983 and 1984 of 4.4 tonnes per hectare works out at 1.73 tonnes per acre), which is about half the yield from conventional techniques, and this, I think, is the best way of looking at it. To make up for this shortfall in production we are at present enjoying a premium for our wheat, and of course we save ourselves the high chemical input cost that the conventional farmer has to bear.

My advice to any young man wanting to start farming organically is, and always has been, to learn first to be a successful conventional farmer. There is so much more to farming than just figures — man management, stockmanship, sifting of advice and control of one's personal affairs — that it is far better to start with a conventional system, where help is available by way of advice, subsidies and chemicals to cover up mistakes and almost unlimited money is being poured into research to combat farming's (largely self-induced) problems. Then, when one has learned how to handle life's problems and to be successful, that is the time to change. For I believe that any successful conventional farmer can be a successful organic farmer.

10

Some advantages of organic farming

Quite apart from the purely business aspects of organic farming, there are several other advantages which seem to develop naturally on land that is being farmed on a sustainable, organic system. It could be argued that some of these apply equally well to a conventional farm if the farmer and/or owner is a true conservationist, and I would not dispute the point. I shall simply point out the changes that have occurred at Rushall since 1970, when we started to convert to organic methods.

To start with, let me deal with a common complaint of observers of the rural scene — field size. In our part of Wiltshire, with its wide open downlands, there have never been any hedges. A few beech belts or clumps have been established over the years, but no systematic division of the land by walls (there is no local stone) or hedges has ever been undertaken. Before the Second World War, when sheep held sway, hurdles made from hazel wood were used to construct sheepfolds over the fields of roots which were grown for these animals. With the increase in fertilizer use and corn production, the fields tended to become bigger and bigger. For instance, the Rushall fields numbered 22, 23, 24 and 26 (see map on pp. 4— 5) had become one block of 133 acres, and, while this made life easy for the combine and the drill, it made life very difficult for the insects, birds and small animals that go to make up the ecology of any area of land. So we have now adopted a system of 'grass hedges', an idea initiated by the Game Conservancy at Fordingbridge, which simply means

A hurdle made from hazel wood

leaving a 10—12 foot strip of grass unploughed when breaking up a ley. This then grows up during the following summer and from then on it provides a refuge and nesting cover for the many beneficial insects and birds which we require on a natural farm. Ladybirds are needed to control aphids when these pests appear on the ears of wheat in the summer. It is difficult to see how they can overwinter in a 100-acre field of wheat that is sprayed continuously, but from these 'hedges', where they are undisturbed from one year's end to the next, they can effectively spread to combat any aphid invasion.

This leads on to another advantage. Most modern farmers, or their advisers, spend hours tramping their fields looking for the first sign of disease or the first aphid. It is not enough to walk them once a year — they have to be walked regularly every week during spring and early summer, and each disease found has to be assessed to determine whether it will be economically sensible to combat it by spraying. Some farmers become nervous wrecks at this time of year as they are persuaded to spend more and more on sprays of one sort or another. The organic farmer, who relies on a healthy soil to produce a healthy plant which can withstand attack by most of the common diseases, can go fishing quite happily because, even if a disease does appear, he may not spray against it. Ignorance is bliss!

Relaxing by the fishing pond

Cleaned wheat

Perhaps one of the more rewarding aspects of organic farming is that you are producing food which people really want to buy. In these days of huge food surpluses it is satisfying to know that there is a ready market for anything produced by organic methods, whether it be vegetables, fruit or arable crops. We have had many people at the mill who have said to us that they have felt very much better since they started eating our bread and flour regularly. Compare this with the absurd situation of intervention buying where unwanted food is produced in ever-increasing quantities, and the sense of satisfaction becomes apparent.

It is somewhat ironic, and certainly not fully understood, that in their quest for whole, pure food the general public are turning more and more to wholemeal flour and bread to introduce more fibre into

The rubbish taken out of wheat by the cleaner

their diet. Yet this wholemeal flour, *unless produced by organic methods*, is likely to contain more harmful chemical residues than white flour. Consider what happens on a conventional farm. The crop of wheat is sprayed, by following the tramlines (see Glossary), right up to the time when the ears of wheat are fully developed, the later sprays being insecticides to control the aphids. If there are any harmful effects from the residues of these sprays they are much more likely to be on the outer surface of each grain of wheat than in the centre, which produces the white flour. So, when commercially grown wheat is made into wholemeal flour, any residues that are on the wheat will be incorporated into the flour. As yet there is insufficient *organically produced* wholemeal flour to satisfy the growing demand for this basic item of many diets, and so people will, perforce, have to go on buying flour with possible residues. This does, however, illustrate how the market could be boosted if organic growing were to increase and were in a position to satisfy the demand.

We have many thatched properties on the farm and these have to be kept in good repair. Quite apart from the fact that untidy thatch looks most unsightly — an eyesore — once the rains start getting through it is not long before the rafters become rotten and have to be replaced. All our roofs are now thatched with what we call 'wheat reed'. To obtain this, a good tall crop of wheat has to be cut with a binder just before the wheat berry itself is hard and ripe. The sheaves then have to be stood up to dry — stooked or 'hiled', as we

Carting sheaves of wheat destined for thatch

say in Wiltshire — before being loaded onto trailers and taken to the barn for stacking. All this is very laborious as it has to be done by hand. Then during the winter, whenever the contractor can be persuaded to come along with his machine, the wheat is combed or 'reeded'. Each sheaf is fed into the machine by hand. One end of the straw is gripped between two belts while the other end is combed — as you would your hair. Halfway along, the position is reversed and the other end is combed. The sheaf then passes out of the machine into a bundler, where it is tied up ready for use by the thatcher. In this way the straw is kept straight and unbroken, and so is a much better material than anything reaped by combine (useless for thatching) or threshed with the old-fashioned threshing machine. Thatchers have a preference for long-strawed varieties (Maris Wigeon is very popular), and they are beginning to appreciate straw which is grown organically. It is easier to work with and the hope and expectation is that it will outlast straw grown with chemicals. Bath University is conducting a long-term experiment into the life of straw as thatch and has purchased a quantity of ours to use alongside other varieties grown by different methods. The old thatchers, who grew up with what was, in effect, organically grown straw, used to say that a roof should last for twenty-five years on the south and west of a house, the sides that take the worst of the weather, and thirty years on the north and east. For those of us around in thirty years time, the Bath experiment should give us the answer!

To fix the straw onto the roof as thatch, wooden spars are used. These are made from hazel wood or from willow, both of which we grow on the farm, and our retired thatcher converts the wood into spars. A finished spar is 2—2½ feet long and pointed at both ends. It is made by splitting a rod of hazel or willow, of about 1-inch diameter, into four, six or eight pieces, depending on the size of the rod. Like most rural crafts, this splitting looks very easy until you come to do it. Watching our retired thatcher carefully, you can see he starts at one end with his very sharp billhook, and then, as he says, follows the grain. I fondly imagined that, if you gave one of these rods a good swipe with the billhook, it would split from end to end with no trouble. It doesn't quite work like that. What happens is that it comes out halfway down and you are left with a useless piece

141

Ernie at work making spars

of wood. Watching Ernie, our ex-thatcher, doing it, you can see him twisting his hook first one way, then the other, so that the split follows right down the centre of the piece of wood he is splitting. 'You must follow the grain,' he says with a certain note of contempt in his voice after seeing me demolish several of his best rods. There

142

is a right and a wrong way to do even the simplest of jobs. Having split the rod into the required number of pieces, both ends are then pointed and placed in a bundle of 500 spars. When the thatcher comes to use them, he twists and bends each spar to form a U or V

Thatching the stables at The Manor, Upavon

Our only remaining thatched barn

with both ends pointed. This is then pushed into the straw and holds it tight. Old-fashioned, yes, but no better way has yet been found.

On an organic farm, which, as I have said, should be a mixed farm with stock to convert the waste products such as straw and to produce income when the land is down to grass, there is plenty of work to be done all the year round. On modern, stockless, 'chemical' farms there is very little to be done once the winter corn has been planted until the spring rush begins. This period may last from November to March and it is difficult to find gainful employment for the tractor-drivers during these months — the farm becomes dormant. As everyone knows, stock has to be tended daily throughout the year, and so, while the emphasis may change each month, as

144

outlined in chapter 3, we are never short of jobs and this leads to less boredom and a more varied working lifestyle for the staff. It is so much better to have too much work to do than to have too little.

At a time when farming's image with the general public is at a pretty low ebb (not entirely the farmers' fault), the organic farmer can plead 'not guilty' to most of the practices that the public has come to regard with aversion — straw-burning, hedge removal, spraying, use of hormones, intensive livestock production, and a certain arrogance towards anyone who questions these practices. Usually the only criticisms we have come from the farming fraternity themselves — we are told we are old-fashioned, that we could not feed the world, that talk of harmful chemical residues is emotional claptrap, that our system could not be viable if generally adopted and that we are benefiting from the chemical fertilizers used for the last twenty-five years; but the general public, be they customers or simply casual observers of the farming scene, are very enthusiastic about the idea of organic farming. It is comforting to know that we do have public opinion on our side and do not need to be constantly on the defensive.

11

The Ministry of Agriculture and others

Reference has been made more than once to the lack of any independent testing of new agro-chemicals by the Ministry of Agriculture. This has always seemed to me a very serious omission, and is paralleled by the Ministry's apparent reluctance to commit itself to any long-term research into organic, sustainable systems of agriculture. Here I shall trace my own efforts to get it involved.

In 1979 I wrote to the newly appointed Minister of Agriculture, the Rt Hon. Mr Peter Walker, and asked him for help in developing organic farming. He passed my letter to his Minister of State, Lord Ferrers, who arranged for me to visit him in the presidential suite at the East of England Show (he was president that year and we were showing a horse in the hunter classes). I spent some twenty minutes with him explaining the lack of Ministry interest and the urgent need for research into sustainable systems of farming for this country, against the day when fossil fuels, on which modern farming depends, became prohibitively expensive. My suggestion was that one of the Ministry's experimental husbandry farms should be directed to adopt an organic system so that ADAS would have some knowledge on which to draw in order to help those farmers who wanted to change over. Failing this, I suggested that each experimental farm should be instructed to convert one field, typical of their area, to an organic system. Finally, if both these suggestions were found unacceptable, I offered our own farm for ADAS to monitor the changes that occur when land is taken off a chemical system and

starts becoming organic. Lord Ferrers listened with considerable interest and promised to let me know what the Ministry could do, while holding out little hope that they would be able to do anything very positive. In due course, I received his reply to the effect that he had decided to accept my offer for ADAS to monitor Rushall, and that he had instructed the Bristol office to carry out the monitoring. While this was obviously something of a disappointment, as it meant the least commitment for the Ministry, it was nevertheless a step in the right direction, and we began to hope.

By its very nature, ADAS is staffed with scientists whose whole lives have been devoted to the domination of nature by science. It is, therefore, easily understandable that they were less than enthusiastic at the thought of working on an organic farm, which seemed to them like regression. So very little was done and nothing worthwhile emerged.

When the Rt Hon. Mr Michael Jopling was appointed Minister of Agriculture I tried again, in a letter dated 13 June 1983:

Dear Minister,

May I first congratulate you on your appointment as Minister of Agriculture and offer my best wishes for a successful and lengthy term of office. It is comforting for the farming fraternity to know that they have a bona fide farmer at the helm.

I realize that at this time there will be very many demands for your attention. There is, however, one that *in the long term* is perhaps the most important and which has been very largely overlooked by previous administrations. I refer to the need for Government backed research into the practice of organic farming.

I made this plea to your predecessor who handed it to Lord Ferrers. He gave me a most sympathetic hearing and directed ADAS to accept my offer to monitor my 1650 acre farm in Wiltshire that is being converted to a wholly organic system — no fertilizers or chemicals of any sort — and which it is hoped

Church bridge, Rushall

to be completely converted by 1990 (1000 acres so far). Unfortunately, however, ADAS were somewhat less than enthusiastic and did little more than pay lip-service to his instructions. Following repeated pressure there has been some slight improvement recently, but no real commitment is evident, largely due they say to lack of funds.

As a farmer yourself you will know that there is a growing number in the industry who view with concern the greatly increased use of powerful chemicals on the land. Also the swelling voice of urban opinion is demanding more natural methods of production of food. This all lends weight to the view that the Government should actively concern itself with building up knowledge on the subject.

My offer is still open — my farm is available for research into the hundreds of questions that need an objective study. To name a few — incidence of disease, soil populations, best rotations for various soils, pest infestations, ecological balances, economic considerations, food quality — these spring to mind and there are many more.

Many millions of pounds are expended annually on research into 'chemical' farming. When one considers that this type of farming is not sustainable in the long term because of the eventual drying up of fossil fuels and that organic farming or, more accurately, a sustainable system of farming, has to all intents and purposes been ignored by government in this country, one realizes how illogical it all is.

Several bodies have sprung up over the last few years — the Soil Association (the oldest), British Organic Farmers, Organic Growers Association, Henry Doubleday Research Foundation, Elm Farm Research Trust — but they all lack the resources to carry out the necessary research. Only Government funds could make this possible. The need is there — it only requires a definite commitment by the Government, of which you are a most important member, to initiate a long-term research project that has been lacking so far. Other countries — the USA, Germany, Holland, Switzerland — are a long way ahead of us which illustrates the urgency of the problem.

Finally may I, with the greatest respect, suggest that this

area presents you, as the new Minister, with your greatest opportunity to benefit the people of this country *in the long term*. In my opinion the need is vital and politically the time is right.

My optimism that a Minister who was himself a farmer would be helpful to the farming community was, unfortunately, misplaced. I believe that Mr Jopling has no sympathy with what we are trying to do and I have received no direct communication from him. He passed me on to his Minister of State, Lord Belstead, who, in a letter to our MP, Mr Charles Morrison, dated 20 February 1984, said, 'I am interested in Mr Wookey's work and if I should find an opportunity to visit his farm later in the year, I would like to do so.' He did so on 18 July 1984 and I believe that it is largely due to his interest that the efforts of the Bristol ADAS have increased so dramatically.

At this stage I should introduce a young lady by the name of Helen Browning, whose research I drew on in chapter 9, as acknowledged there. She is the daughter of a great friend of ours and had been helping her father with the lambing for two or three years on his farm near Swindon. In May 1984, when she was taking a degree course at Harper Adams College in Newport, Shropshire, she rang me one day and explained that her course required her to do a year's practical work on a farm and to write up a report on what she had done. Her father, who is one of the best 'chemical' farmers in Wiltshire, had realized that she was becoming increasingly worried at the rapid increase in the use of chemicals on the land, and so had pointed her in our direction. Thus, when she rang and asked to come and see me I was not totally unprepared. She said that, because of her worries about the way modern farming was going, she wanted to find out more about the organic approach and would be grateful if she could work on our farm for the year. After a few days thought I suddenly had one of those flashes of brilliance that occur so rarely! ADAS was still dragging its feet a little, largely because of shortage of staff and cash, so I felt this might be a chance to help. If Helen could work for ADAS for a year and do the research work on our farm that it had found it couldn't manage unaided, then the work would get done and Helen, with the laboratory resources of the

Bristol ADAS to support her, could prepare an original and worthwhile report for her degree. So I rang Dr Burgess in London, who with a Dr Little had already been down to the farm, and suggested that Helen could be the eyes and legs for the ADAS in Bristol. He told me that the Service did have 'studentships' for this type of work, but that they were very few and there were thousands of applications, so, while it was possible for Helen to be given one, it was highly unlikely. He would investigate and let me know and meanwhile I should mention the matter to Lord Belstead when he came on his visit. I was delighted to hear from Helen about a month later that the ADAS had agreed to take her on for the year and that she hoped the project would prove 'as interesting and helpful as I believe it should be'.

The work that Helen undertook, under the direction of Mr Unwin of the Bristol ADAS was very extensive and included the following subjects, which I set out with the summary produced by Helen Browning and the Bristol ADAS. The work was carried out on twelve of our organic fields and five on neighbouring farms where chemicals are used.

1 Weed studies

Weed populations in late June were similar to previous seasons except for a conventional field where herbicides were ineffective. Nitrogen uptake studies indicated that in late April 25 per cent of the total nitrogen uptake in organic fields was in the weed flora. By late June the corresponding value was 5 per cent. Hand-weeding studies in two organic fields failed to show a benefit from removing weeds in late April and again in late June.

2 Cereal-crop monitoring

A total of ten cereal fields (seven organic and three conventional) were monitored in 1984—5. Severe winter weather conditions were unfavourable, nitrogen uptakes on the thin chalk soils were low and the crops yielded poorly. The average of five such organic fields was 3.28 tonnes per hectare (26 cwt per acre). In contrast, two first wheats in the valley yielded in excess of 5

A pyramid orchid

tonnes per hectare (40 cwt per acre). *Septoria tritici* was present in two organic fields at levels which are likely to have reduced yield.

3 Grass-yield assessments

Grass yields were measured on five organic grazing fields and two cut for hay. Yields of first-year leys in the process of conversion into the organic system were low. However, a second-year ley in this situation and two second-year leys in established organic fields yielded an average of 8.6 tonnes per hectare dry matter. Hay yields were 5.8 and 6.8 tonnes per hectare although part of this same field yielded only 3.1 tonnes per hectare apparently due to phosphorous deficiency. A conventional field cut for silage probably yielded around 11 tonnes per hectare with additional aftermath grazing which was not assessed.

4 Winter wheat: milling variety trial

Variety trials were undertaken in two organically farmed fields with different previous cropping. Severe weather reduced plant populations over winter generally by 40–60 per cent. The effect was particularly severe at one site thus preventing comparisons between residual fertility differences. The newer varieties Avalon, Mission and Brimstone outyielded Copain and Maris Wigeon at the other site. Quality tests for baking indicated that the Mission, Brimstone and Maris Wigeon were comparable and above average for conventional farms in 1985.

Wheat in store

5   Soil mineral-nitrogen study

Soil samples were taken at regular intervals throughout the year. Mineral-nitrogen levels and inferred levels of loss were similar for organic and conventional fields. Minimum levels in the organic fields occurred in early April and were very low at this time. High mineral-nitrogen values were recorded after leys, red clover and beans on the organic fields. There were apparently large losses over winter but three of the four following wheat crops performed well on the residual fertility.

6   Phosphate and potassium on Manor Farm, Rushall

After five years of irregular monitoring of P [phosphorus] and K [potassium] levels of selected fields it is apparent that few serious P and K deficiencies occur at present on the farm. The main exception to this is the P levels in Charlton North Field East (no. 17), and Cuckoo Pen (no. 22), which have very low levels (index 0) since monitoring began in November 1979. Indices are generally adequate or high.

7   Earthworm survey of organically and conventionally farmed fields

A total of twenty-two fields under both organic and conventional management were sampled for earthworms between 30 October 1984 and 28 November 1984 using formaldehyde expulsion techniques. Initial examination of collected data on earthworm numbers, biomass and species indicates significant differences between the organic and conventional farming systems examined; however, a large number of factors have been shown to affect earthworm populations and not all the differences can be attributed to the use of chemicals.

8 Financial assessment of an organic farming system

On a 1:1 grass:cereal rotation the overall margin at Manor Farm can be calculated at £377 per hectare, which is comparable with conventional farms operating the same balance of enterprises.

These eight projects represent a prodigious amount of work that was undertaken, almost single-handed, by Helen Browning. She was supported by Roger Unwin of the Bristol ADAS, but the sheer volume of work she devoted to these projects during the year reflects her worry at the direction of modern farming methods. It is work that should have been started by ADAS in 1979, when I first approached Peter Walker, and it is work that should be continued if sufficient data is to be collected to enable the ADAS advisers to be of any help to farmers seeking guidance on organic methods. Since Helen Browning went back to college in September 1985 another student has been taken on for six months (April to September 1986), after which time no more Ministry money is to be made available for this project. When one considers that on 28 February 1986, in reply to a question on Government funding for organic research, the Minister stated, 'Given competing claims for limited research resources, such a contribution could not be justified at public expense', and that the same year the Ministry decided to drop its plans to introduce an organic unit on one of its experimental husbandry farms, it becomes clear that the Ministry has still not realized the importance of this type of work. Both Mr Walker and Mr Jopling have let slip the opportunity to do something worthwhile, and one wonders when we shall have a Minister with the vision to see that the future of mankind depends on the establishment throughout the world of sustainable systems of agriculture to replace the exploitation of the earth's resources for short-term gain. (See also the proposals presented in chapter 12.)

I am not sure whether the Ministry's lack of response is a reflection of its assessment of the feelings of the general public or whether it really does believe that any money spent on organic systems would be money wasted. On both counts I think it is wrong, if our own experience is anything to go by. When we started this enterprise in 1970 I admit that there was very little interest, either from the general public or from fellow farmers. We had to wait three years for the first fields to be ready to produce the wholemeal flour on which we based the whole enterprise, so it was not until 1974—5 that people began to take notice. Since then, as our visitors book shows, there has been an ever-growing number of people, both interested parties and professional farmers, who have been to the mill and looked round the farm. As already noted, we have had visitors from over thirty countries, in all parts of the world, and the interest these people have all shown is ample proof to us, if not the Ministry, that there is a definite feeling of unease at the way modern agriculture is developing. The signs that we put up from time to time — 'No Chemicals since 1972' — attract enormous interest and have been the inspiration for several radio and television programmes. Again, when we first started selling wholemeal flour, we were excited by every 25-kg bag that we sold; now we are disappointed if we do not shift 10 tonnes a week.

Of course, with a new venture such as this we have had some worrying moments! One I recall very vividly. I happened to be in the mill one morning when a customer walked in — quite an event in the early days. The man running our mill then was an ex-soldier called Sid Bird — 'Dickie', naturally, to the others on the farm — and we started talking to the visitor, who turned out to be an employee of one of the milling giants in the country. He bought a 25-kg bag and Sid and I felt we had done a good morning's work, selling a bag of flour to a representative of such a large miller.

About three weeks later I was in the mill again when in walked the same man. This time he was carrying our bag of flour, by now half empty, and we could see at once that he was far from happy. He put his hand into the bag and pulled out a very dead mouse. He sounded off quite well for a few minutes, saying how disgraceful it was to sell him a bag of flour with a dead mouse in it, and how upset his wife had been to find the mouse in the bag. I was struck completely dumb — I didn't know what to say or do, but I do remember thinking how delighted this man's firm would be to take us to court and that it would probably be the end of our business. Not so Sid! With all his years of experience in the sergeant's mess coming to his aid, he took

David Fuller in the bakehouse

the dead mouse, examined it minutely and then handed it back with the words, 'It's not one of ours, sir.'

On another occasion I had a letter from an irate customer enclosing a cellophane packet containing what could only be a piece of well-chewed gum. This had appeared when his wife was using our flour to bake some bread and rather naturally they had been somewhat annoyed. I went over to the mill and demanded to know if Sid or his successor, David Fuller (they were working together at the time), used chewing gum. Both denied it, so we started working out how it could possibly have turned up in a bag of our flour. Eventually we decided that it must have been stuck on the inside of the flour spout in the flour house by one of the many children we show round the mill. I gather that when children get bored with chewing a piece of gum it is their habit to stick it on the underside of their desk or any other handy surface. In this case it must have been the flour spout, so now we do not allow children into the flour house. We live and learn!

Not all incidents are quite so discouraging. I remember one evening I was speaking at the Piscatorial Society annual dinner and our organic enterprise was mentioned in the chairman's opening remarks. Quite unknown to me, Lord Penney, the leading atomic scientist of the day, was in the audience, and at the end of the dinner he came up and said how interested he was to hear about what we were trying to do at Rushall, because, he said 'in what you are doing I see some hope for the future, but in modern farming practice I see none'. This has been a source of great encouragement to us when things haven't been going too well, and on the strength of that remark I approached him to see if he would be prepared to write a foreword to this book. The reader will already know that he most kindly consented to do so.

Another encouraging aspect of our adventure has been the interest of the younger generation in what we are doing. I have found that farmers of my own generation are usually quite content to continue as they are — they have established an easy and profitable system with the use of chemicals, so why change? The younger ones, often more idealistic than practical, but fired with the enthusiasm of youth, are the ones who ask the searching questions and seem keener to learn than to ridicule. In the last few months, two sons of farmer friends of ours have said to me independently that they believe agriculture should be changing direction, and this is most encouraging for the long-term future. People of my generation will not be around to suffer from the effects of the modern throw-away society — the next generation and their children will be the ones to reap the penalties of our short-sighted methods, and maybe it is for this reason that the young are the ones who seem to be most concerned.

12

The future

In any debate on the future direction of farming in Britain there are several fundamental facts that should be borne in mind. While they are all, to a greater or lesser extent, interdependent, it is necessary to appreciate the part each one plays in the process of food production in Britain, so that we can establish the needs for the future. (Similar concerns, of course, apply in other EEC countries and elsewhere in the world.)

These fundamental facts include the following.

1　Consumer preference. There is a growing demand for food that has been produced without chemicals. Whether the fear of chemical contamination is justified, as I believe, or just a neurosis, the fact remains that there is a ready sale for this type of produce.

2　Conventional farming depends on two things: high inputs of fertilizer and chemicals derived largely from non-renewable sources, and a high level of financial support from the EEC's Common Agricultural Policy and the Government. (It has been estimated that it cost £1,500 million just to store the Community surpluses in 1985.)

3　Much of the output of high-input farming is bought by the bag — bagged fertilizer in the case of crops, and bagged fertilizer and cake for milk and meat production.

4　There is increasing evidence that the side-effects of chemicals can be harmful. Take any farming paper and you

The control panel in Rushall Mill

will find a reference to some new problem — it could be toxicity, persistence or lack of selectivity.

It is against this background that we must try to find sensible solutions. The more one thinks about it, the more critical and urgent the problem seems to become. There are already signs that nature may be fighting back against the domination exerted by scientists over the last half-century, a minute span of time in natural systems. I have already mentioned the fear that apparently unaffected plants may succumb to repeated bombardment by apparently benign chemicals. The widely reported case of 'Superbugs versus wonder-drugs' (BBC2 *Horizon*, 3 February 1986) drew attention to the contamination of half the hospitals in Melbourne by a strain of *staphylococcus* which was resistant to almost all known antibiotics. It suggested that doctors were now facing the possibility that the benefits of the antibiotics revolution have been squandered by over-subscription and that, far from being controlled by drugs, future generations of microbes may actually thrive on them. In the arable sector of farming we are seeing new strains of disease appearing every year, and we hear of organisms becoming immune to certain hitherto effective remedies.

Another real worry is with the nitrate levels in our water supplies. As in all these things, it is very difficult for the layman to know what to believe when one expert says one thing and another something quite different — frequently the opposite. In many parts of Britain, water supplies already have a nitrate level above the EEC recommendation of 45—50 parts per million, and, as we are told that it takes between twenty and eighty years for the nitrogen used today to reach the water tables which supply our drinking water, we are in danger of creating serious problems for our grandchildren.

We also have to contend with vested interests. Many millions of pounds are spent on marketing agro-chemicals, a huge industry which sells sprays to the value of some £345 million yearly (British Agro-chemical Association, *Annual Report and Handbook 1984/85*). In addition, vast quantities of artificial fertilizer are sold every year, and, while everyone would agree that it has done a wonderful job in increasing production — indeed, it was thanks largely to such fertilizer that Britain managed to feed itself during the Second World War — the time has surely come to question whether all this money is being spent in the best interests of the public. Surely some part of it could be diverted to the huge areas of research that need attention and which, despite their importance to the future of the human race, are currently receiving none, or very little. Whether the resources are made available is, of course, ultimately a political decision, and politicians, by their very nature, are not concerned with planning for the long term. There is, however, no excuse for the Civil Servants, who should be thinking and planning far ahead and convincing their masters that they must act now before it is too late.

So what should the Ministry of Agriculture be doing? On the one hand, as already argued (see chapter 11), it should be examining the

The future

possibilities of adopting sustainable systems on our farms, and, on the other, it should be exercising its responsibility as Ministry of Food by establishing a workable system for licensing organically grown produce to protect those people who are demanding it.

A loaf in the form of a sheaf of wheat, baked for Harvest Festival every year

As this is at the very heart of the matter, we should examine this proposal in more detail. The first thing to establish is what 'organically produced food' is. At the moment there is absolutely no Ministry definition of such food. The only guarantee that the consumer has in buying food that claims to be organically produced is the symbol scheme of the Soil Association, who license the use of their symbol to those producers who grow to the Soil Association Standard, set by the British Organic Standards Committee (see chapter 4). This committee is chaired by Laurence Woodward of Elm Farm Research Centre, and composed of representatives of all the organic organizations in Britain. These are the organizations concerned:

> The Bio-Dynamic Agricultural Association, for followers of Rudolf Steiner
>
> British Organic Farmers, an organization for organic farmers and other interested parties
>
> Elm Farm Research Centre, a trust set up to conduct research into organic farming
>
> The Farm and Food Society, concerned with animal welfare
>
> The Henry Doubleday Research Association, for organic gardeners
>
> The International Federation of Organic Agriculture Movements
>
> The International Institute of Biological Husbandry, whose aim is to promote scientific work in organic agriculture
>
> Organic Farmers and Growers Ltd, a commercial company trading in organic produce (mainly cereals)
>
> The Organic Growers Association, for organic growers and other interested parties
>
> The Soil Association, which seeks to further the concept of holistic agriculture

The Standards Committee freely admits that there are short-comings to the Soil Association Standard, but it is the only document on the subject for British producers. There is nothing from the Ministry and so it would be very difficult to take action against anyone found selling food as organic which did not meet the Standard. What is urgently needed is a definition of organic produce, an official set of standards, a licensing system and an inspectorate all organized by the Ministry. Only in this way can consumers be certain that what they buy in all good faith as organic is really genuine. It is a big undertaking, far too big to be undertaken properly by small voluntary organizations, and time is slipping by. As the demand grows (and it is growing very rapidly, to judge from our own sales), so the temptation to label produce 'organic' to attract a higher price will become irresistible to some who do not observe the Standard and the consumer will be cheated. This danger was acknowledged by the Minister, Mr Jopling, when he visited the Soil Association stand at the 1986 Royal Show. He said then that something should be done to prevent the 'cowboys' from jumping on the bandwagon, so that consumers could be sure what they were buying. He mentioned the word 'cowboy' several times in my hearing, but, as with all such visits, time was pressing and he had to rush off to another stand, so whether he was serious or was just trying to placate a minority group only time will tell.

I believe that there is a place for organic farming in the future agriculture of Britain, the EEC, and indeed the world. Let us consider for a moment the present position. We, in the developed world, are producing large quantities of unwanted food, much of which is unsuitable for feeding the starving peoples of the under-developed countries. A large proportion of the £650 million worth of grain at present (1986) in British intervention stores is unsuitable for milling as it is what is called 'feed wheat'. This is a type of high-yielding wheat for animal feed that responds well, having been bred for that purpose, to high inputs of chemical fertilizers, and it costs many millions of pounds to dispose of it on the world markets (see chapter 9). Milk and beef are also in over-supply — quotas have been introduced to try to reduce the flow of milk into intervention stores, and beef production is still increasing. All this is costing the taxpayer thousands of millions of pounds — £1,500 million just to *store* the surpluses in 1985 alone. Butter, which costs some £2,000 per tonne to put into store, is being sold for £200 a tonne to reduce the mountain, as it will not keep indefinitely. And what is it going for? Cattle food. What a ridiculous situation!

All governments in the developed countries support their agriculture; this, whether we like it or not, is a fact of life. The question that confronts our policy-makers in this time of surplus is

A Cambridge ring roller at work

whether they are using our resources to the best advantage. It is, I think, recognized that in the EEC things have got out of control, and already ideas are being put forward to alleviate the problem. One such idea is 'set-aside' payments — to pay a farmer *not* to plant his land. The proposal is that some, possibly 10 per cent, of all land should be fallowed for a year and the farmer would be paid £75—80 per acre for complying. This is to my mind a very negative proposal, but it does establish the principle that the government should pay farmers directly rather than via the intervention system. Another idea is to plant trees on arable land and to pay the farmer £60 an acre for doing this — £30 capitalized to start the plantations and the other £30 an acre to be paid yearly as income. The drawback to this idea is that farmers on the better arable land, chiefly in the eastern counties of England, where nitrate levels in water are already very high, would not consider this a sufficient income compared to corn-growing, with the result that only low-quality land would be taken out of production while the better land would be farmed more intensively. When one realizes that some 4—4.5 million tonnes of soluble nitrogen fertilizer are used in Britain annually and that this would probably be concentrated even more heavily in the eastern counties should a tree-planting policy be introduced, it is easy to imagine the potential damage to future water supplies.

It is in this context that I think that organic farming should be seriously considered by our policy-makers as a means of reducing the surpluses of unwanted food, and of ensuring a well-farmed countryside. Recall, for example, the following points, all of which argue in favour.

1 The EEC governments are considering paying farmers £75—80 an acre to produce nothing.
2 Conversion to organic farming means reduced yields overall and much reduced yields initially.
3 Organic farming means mixed farming — stock as well as crops.
4 Organically produced food is in ever-increasing demand.
5 Organic farming is sustainable.
6 The general population is becoming restive at the direction of modern agriculture.
7 Conventional farming practices mean using up finite resources to produce commodities which are not wanted.

Surely it is not unreasonable to suggest that a direct financial encouragement to farmers to turn over to organic systems of farming should be introduced. It makes more sense to use money to help farmers produce food that is required than to pay them under the 'set aside' policy for producing nothing. And, when one takes into account the cost to the health service of treating food-related problems (cardiovascular disease, diverticular disease, cancer of the gut and bladder, obesity and strokes, gallstones and dental disease, allergies from food additives and so on), and the cost to the water authorities of water purification when nitrate levels become dangerously high, the proposal seems all the more logical.

There should be an 'organic incentive' payable to all farmers who adopt an organic system. At the moment organic farmers receive no government assistance by way of subsidy, research and development, or advice, while conventional farmers have all this help. What I envisage is a system of incentive payments starting, say, at £80 per acre in year 1 and falling to £50 per acre in year 4, then continuing at that level indefinitely. This would reflect the difficulties we experienced in the first few years of conversion and help to offset some of the mistakes that are bound to be made. It would also make it possible for people with high rents and borrowings to consider an organic approach, which at present, owing to the lack of government help, is an option denied them. Obviously the details would have to be worked out, such as how to handle the case of a farmer who suddenly opted to revert to conventional farming and so by rights ought to repay at least some of the previous assistance he had received, but I cannot see that this would present any more difficulties than are encountered with the hill-cow subsidy or the intervention system itself. Among the many advantages of an incentive system would be the following.

1   It would enable many farmers who would like to do so to change to a sustainable system.
2   It would save a lot of money — no surpluses.
3   It would help to prevent the build-up of nitrates in our water supplies.
4   Our finite resources — fossil fuels — would last longer, to be used in times of emergency.
5   Our countryside would all be farmed — no acres of fallowed land that could support no wildlife.
6   Mixed farming would return, so enhancing the countryside

Above: High summer at Rushall: a marbled white on scabious

Right: All the species of our long ley mixture (see table 6.3, p. 95)

  — more stock, more hedges and windbreaks, more wildlife and wild flowers, and so on.

7 Everything produced would be in demand thanks to increasing public awareness of the dangers of modern agriculture.

8 The policy would be attractive to farmers of all grades of land — those on the best land could see a return comparable to that on chemical systems.

What should happen is this: the Ministry of Agriculture, in consultation with all interested parties, should first define organic produce and then formulate a set of standards to apply to organic

production, giving these standards the force of law. It would then need to license producers to enable them to sell their products as organic on the open market, and to do this it would need an inspectorate both to license and to police the scheme and to administer the 'organic incentive' proposed above. In addition to all this, the Ministry should develop the ability to advise farmers who wish to convert to organic systems. At present there are no ADAS staff trained in these skills, so would-be organic farmers have to address their enquiries to British Organic Farmers or individuals who are already farming organically.

To sum up, I believe that the Ministry should be channelling the money now being used to develop new chemicals into research into sustainable systems of farming that will (1) produce the type of food people want, and (2) become independent of high inputs of finite resources. Surely, in all logic, this is what is urgently needed, and, if this book can help to point our political masters in this direction, then it will have achieved its purpose.

I leave the last word with Frederick W. H. Myers (1843–1901):

Dreamers of dreams? We take the taunt with gladness,
Knowing that God, beyond the years we see,
Has brought the dreams that count with man for madness
Into the fabric of the world to be.

Meadow crane's bill

Appendix A

Rushall 1986: handout for visitors

1 Cropping (acres)

Winter wheat (Maris Wigeon, Flanders; Mission, Brimstone)	130
Winter/spring wheat (redrilled after winter kill)	273
Winter oats	13
Winter/spring oats (redrilled after winter kill)	30
Spring wheat (after kale, etc.)	75
Long ley	671
Short ley (Italian ryegrass/red clover)	71
Roots (mangolds, swedes, kale)	43
Permanent grass	230
Woodland, tracks, etc.	114

	1650
Rough grazing with 60 acres of barley on Ministry of Defence Larkhill Ranges	400
	2050

2   Stock

(a) *Beef cattle*. All Friesian × Hereford Steers for sale as stores

1 — 2 year olds	147
6 — 12 mths	271
	418

(b) *Sheep*. Joint sheep enterprise with John Dewey, who owns the sheep, provides the labour and equipment and pays headage payment for year-round keep.

Breeding ewes (Mules)	700
Lambs	1225
	1925

Headage payment: ewes 25p a week, lambs 11p a week from date six weeks after lambing starts until 1 October, when they are classed as adults.

(c) *Horses*. 10 good quality colt foals bought each autumn (by Hunter Improvement Society and/or Thoroughbred horses), reared on and sold at biennial sales in June — last sale on 12 June 1985, when 23 were sold.

Named with letter of year — 1985-born foals have names starting with letter K.

Ks (1 year old)	10
Js (2 years old)	10
Is (3 years old)	7
Hs (4 years old)	4
Other ages	6
	37

3 Ley mixture

[Details given in chapter 6 as table 6.3, p. 95]

4 Staff

The estate employs a total of 16, comprising:

    1 manager — Richard Fry
    1 miller/baker
    8 tractor-drivers
    1 keeper/estate policeman/vermin-controller
    1 girl groom
    1 maintenance man (30 cottages, plus buildings and water supply)
    1 part-time pensioner on grass-cutting
    1 part-time help in the mill — the miller's wife
    1 part-time secretary — Marilyn Treverrow

5 The mill

Modernized in 1973. Set up to take harvest grain through to flour and bread. Comprises:

    15 tonnes/hour Alvan blanch drier
    30 tonnes/hour pre-cleaner
    3 tonnes/hour Eureka cleaner
    36-inch Simon Barron stone mill
    8 bins at 32 tonnes
    8 bins at 100 tonnes
    Floor storage for 800 tonnes
    Floor storage for 1,000 tonnes

Capable of drying, cleaning, cake-making, flour grinding and bagging, oat-rolling and barley-pearling.

Wheat for milling is blended in proportion of varieties available, into one bin. From there it goes through a polisher, cleaner and indented cylinder to the holding bin above the stone mill. After milling it goes through a Le Coq sieve, to remove any unground bits, to a 3-ton hopper above the semi-automatic bagger/weigher.

6   Mill prices (all ex-mill)

Flour in

1 tonne lots	£230 per tonne
0.5 tonne lots	£250 per tonne
25 kg bags	£7 per bag (£280 per tonne)
12.5 kg bags	£4 per bag (£320 per tonne)
1.5 kg bags	60p per bag (£400 per tonne)
Milling wheat	£200 per tonne in bulk

7   Yield

Wheat average yield:

1.8 tonnes per acre in 1984
1.4 tonnes per acre in 1985

Appendix B

Wild flowers on Rushall Farm, summer 1984

Grass verges

Privet	*Ligustrum vulgare*
Nettle	*Urtica dioica*
Knotgrass	*Polygonum aviculare*
Broad-leaved dock	*Rumex obtusifolius*
Common mouse-ear	*Cerastium fontanum*
Red campion	*Silene dioica*
Bladder campion	*Silene vulgaris*
Creeping buttercup	*Ranunculus repens*
Hedge mustard	*Sisymbrium officinale*
Hairy bittercress	*Cardamine hirsuta*
Shepherd's purse	*Capsella bursa-pastoris*
Lesser swinecress	*Coronopus didymus*
Wild mignonette	*Reseda lutea*
Agrimony	*Agrimonia eupatoria*
Dog rose	*Rosa canina*
Bramble	*Rubus fruticosus*
Herb bennet	*Geum urbanum*
Silverweed	*Potentilla anserina*
Tufted vetch	*Vicia cracca*
Meadow vetchling	*Lathyrus pratensis*
Sainfoin	*Onobrychis viciifolia*
Rest-harrow	*Ononis repens*

Kidney vetch	*Anthyllis vulneraria*
Common melilot	*Melilotus officinalis*
Lucerne	*Medicago sativa*
Bird's foot trefoil	*Lotus corniculatus*
Black medick	*Medicago lupulina*
Hop trefoil	*Trifolium campestre*
Red clover	*Trifolium pratense*
White clover	*Trifolium repens*
Meadow crane's-bill	*Geranium pratense*
Herb Robert	*Geranium robertianum*
Common mallow	*Malva sylvestris*
Perforate St John's wort	*Hypericum perforatum*
Common rock-rose	*Helianthemum nummularium*
White bryony	*Bryonia cretica*
Ivy	*Hedera helix*
Wild carrot	*Daucus carota*
Ground elder	*Aegopodium podagraria*
Hogweed	*Heracleum sphondylium*
Wild parsnip	*Pastinaca sativa*
Yellow-wort	*Blackstonia perfoliata*
Hedge bindweed	*Calystegia sepium*
Field bindweed	*Convolvulus arvensis*
Crosswort	*Gatium cruciata*
Lady's bedstraw	*Galium verum*
Cleavers, goose-grass	*Galium aparine*
Self-heal	*Prunella vulgaris*
White dead-nettle	*Lamium album*
Hedge woundwort	*Stachys sylvatica*
Bittersweet	*Solanum dulcamara*
Great plantain	*Plantago major*
Ribwort plantain	*Plantago lanceolata*
Elder	*Sambucus nigra*
Field scabious	*Knautia arvensis*
Small scabious	*Scabiosa columbaria*
Teasel	*Dipsacus fullonum*
Pineapple weed	*Matricaria suavedens*
Daisy	*Bellis perennis*
Yarrow	*Achillea millefolium*

Ox-eye daisy	*Leucanthemum vulgare*
Ragwort	*Senecio jacobaea*
Lesser burdock	*Arctium minus*
Spear thistle	*Cirsium vulgare*
Lesser knapweed	*Centaurea nigra*
Greater knapweed	*Centaurea scabiosa*
Goat's-beard	*Tragopogon pratensis*
Chicory	*Cichorium intybus*
Sow thistle	*Sonchus oleraceus*
Wall lettuce	*Mycelis muralis*
Dandelion	*Taraxacum vulgaris*
Common cat's-ear	*Hypochaeris radicato*
Rough hawkbit	*Leontodon hispidus*
Pyramidal orchid	*Anacamptis pyramidalis*
Common spotted orchid	*Dactylorhiza fuchsii*
Fragrant orchid	*Gymnadenia conopsea*

Cornfields

Black bindweed	*Polygonum convolvulus*
Field poppy	*Papaver rhoeas*
Charlock	*Sinapis arvensis*
Sun spurge	*Euphorbia helioscopia*
Field pansy	*Viola arvensis*
Scarlet pimpernel	*Anagallis arvensis*
Henbit	*Lamium amplexicaule*
Black nightshade	*Solanum nigrum*
Common toadflax	*Linaria vulgaris*
Field speedwell	*Veronica persica*
Welted thistle	*Carduus acanthoides*
Corn sow thistle	*Sonchus arvenis*
Common forget-me-not	*Myosotis arvenis*

Appendix B

By water

Yellow water-lily	*Nuphar lutea*
Meadowsweet	*Filipendula ulmaria*
Great willow-herb	*Epilobium hirsutum*
Purple loosestrife	*Lythrum salicaria*
Common comfrey	*Symphytum officinale*
Hemp agrimony	*Eupatorium cannabinum*

Appendix C

Soil Association List of recommended, permitted, restricted and prohibited substances and practices

Manure Management

The management of animal manures, crop residues and off-farm organic material should aim to achieve maximum recycling of nutrients with minimum losses. Manure from all types of livestock carried on the farm should be utilized. It is a principle of biological husbandry that stock should be maintained under extensive conditions (see chapter 4, 'Livestock Husbandry Standards'). When importing manures onto the farm, every effort should be made to obtain products from these sources.

All bought-in manures must be approved by the Symbol Committee and must be composted before use. They should not, in general, form the basis of a manurial programme, but should be adjuncts. Some exceptions, however, are acceptable.

Recommended	Own farm FYM either composted or stockpiled undercover
	Aerated slurry or urine
Permitted	Own farm FYM either stockpiled outdoors or fresh
	Own farm tank or lagoon-stored non-aerated slurry or urine
Restricted	Composted off farm manures and wastes — after analysis with Symbol Committee's permission

Bought-in aerated animal slurry

Bought-in worm composts — after analysis and
with Symbol Committee's permission

Symbol approved manures and composts

Mushroom composts — symbol approved only

Municipal compost — symbol approved only

Sewage sludge — not more than one year in
three, only on crops not for human
consumption, after analysis and with Symbol
Committee's permission

Prohibited   Bought-in manures that are not composted

Unaerated bought-in slurry

Manures from ethically unacceptable livestock
systems (from 1 January, 1989). These are
defined as: battery system and broiler poultry
units; indoor tethered sow breeding units; other
systems where stock are not freely allowed to
turn through 360°, where they are permanently
in the dark, or are kept without bedding

Care must also be taken when spreading manures and compost to
avoid leaching of nutrients and environmental pollution. Excessive
manuring must be avoided.

The Soil Association reserves the right to introduce limitations on
tonnage of manurial applications in any one year, in order to:

(a) reduce the risk of nitrate contamination of water courses,
particularly in high risk areas, and

(b) ensure that nitrate levels in Symbol foodstuffs do not exceed
the recognized safety limits that are already legally enforceable
in some EEC countries.

Heavy Metals

Heavy metal levels must not exceed:

	in soil *(mg/kg)*	*in manures* *(mg/kg)*
Zinc	150	1,000
Copper	50	400
Nickel	50	100
Cadmium	2	10
Lead	100	250
Mercury	1	2

Annual rates of addition of heavy metals to the soil by means of manure will also be subject to certain limits. Any analyses required will be at the applicant's expense.

Composting

Composting is defined as a process of aerobic fermentation. A substantial temperature increase must occur within the heap (up to 70° C), and the materials must be turned once to allow a second increase in temperature, unless the material is aerated by other means. The heap should be maintained for at least three months.

Mineral Fertilizers

Mineral fertilizers should be regarded as a supplement to, and not a replacement for, nutrient recycling within the farm. A slow and balanced uptake of nutrients by the plant must be aimed for. In general, only fertilizers that release nutrients through an intermediate process, such as weathering or the activity of soil organisms, are allowed.

Single mineral or naturally occurring compounds are recommended. Compounded organic fertilizers must be specifically approved by the Symbol Committee before use.

Restricted use of some highly soluble nutrients, either naturally occurring or recycled organic material (e.g. blood meal) will be allowed in certain situations. In the absence of more acceptable inputs, restricted use of soluble fertilizers to treat severe potassium or trace element deficiencies is allowed with the specific approval of the Symbol Committee.

Permitted	Rock phosphate
	Feldspar
	Magnesium limestone (dolomite)
	Calcium sulphate (gypsum)
	Ground chalk
	Limestone
	Seaweed
	Unadulterated seaweed foliar sprays
	Calcified seaweed
	Basic slag
	Rock potash
	Bonemeal
	Fish meal, hoof and horn meals
	Woodash
Restricted	Dried blood — in protected cropping, otherwise only with Symbol Committee's approval
	Wool shoddy, hop waste
	Leather meal
	Sulphate of potash
	Sulphate of potash — magnesium
	Kieserite
	Borax
	Epsom salts
	Symbol approved organic fertilizers
	Aluminium phosphates — neutral/alkaline soils only
Prohibited	All other mineral fertilizers, including:
	Nitrochalk
	Chilean nitrate
	Urea

Muriate of potash
Kainit
Slaked lime
Quicklime
Proprietary organic fertilizers without
    Symbol approval

Weed Control

Weeds can be seen as indicators of soil fertility and management
practice in a farming system. Thus weed population can be altered
as a direct result of changes in farming practice. Many 'weed problems'
are caused by imbalances in the design and management of the
farming system. Primary control must therefore be approached in a
whole system rather than in a crop- or field-specific way. The
objective of weed control within organic production is to suppress
populations rather than eliminate them.

Weed control can be effected within organic systems by giving
attention to the following practices: rotation design, manure
management, fertilizations, varieties, seed rates, utilization of green
manures, pre-seeding cultivations and sowing dates. Direct intervention
in the growing crop can thus be minimised and restricted to mechanical,
hand or thermal operations.

Recommended   Balanced rotation (see above)
               Composting
               Slurry aeration
               Pre-sowing cultivations
               Sowing dates (e.g. stale seed bed)
               Pre-germination, propagation, transplanting
               Higher seed rates
               Variety selection
               Undersowing
               Utilization of green manures
               Raised beds
               Biodegradeable mulches
               Mixed stocking

187

	Hygiene — field, machinery and seed
	Pre- and post-emergence mechanical operations
	(e.g. hoeing, harrowing, topping, hand labour).
Permitted	Pre- and post-emergence thermal operations
	Plastic mulches
Prohibited	All chemical and hormone herbicides, within the crop, at the edge of fields, within or below hedgegrows, headlands and pathways.

Pest and Disease Control

Pest and disease control in organic agriculture are primarily preventative rather than curative. At present, because of the lack of technical development, currently available remedies for pest or disease control are often inefficient, expensive or both. In addition to good husbandry and hygiene, the key factors of pest and disease control are:

(a) balanced rotational cropping to break the pest and disease cycles;
(b) balanced supply of plant nutrients;
(c) the creation of an ecosystem within and around the crop which encourages predators, utilizing, where appropriate, hedgerows or mixed plant breaks within fields, companion planting, undersowing and mixed cropping;
(d) the use of resistant varieties and strategic planting dates.

Recommended	Balanced rotation (see above)
	Creation of a diverse, predator-encouraging ecology both within and around the crop (including companion planting, mixed cropping, undersowing)
	Resistant varieties
	Strategic planting dates
	Balanced nutrient supply (see above under 'Manure Management' and 'Mineral Fertilizers').

Permitted   Mechanical controls using traps, barriers and
         sound
        Pheremones
        Herbal spray, homoeopathic and biodynamic
         preparations
        Waterglass (sodium silicate)
        Bicarbonate of soda
        Soft soap
        Steam sterilization
        Biological control with naturally occurring
         organisms
        Conventionally grown seed — recleaned only

Restricted   (a) *Routine* use of the following substances must
        have the Symbol Committee's approval:
        Pyrethrum
        Derris
        Quassia
        Copper
        Sulphur
        Metaldehyde tape — only on non-cropping
         areas
        (b) *Any* use of the following only allowed with
        the Symbol Committee's approval:
        Non-mercurial seed dressing
        Traps containing prohibited insecticides
        Potassium permanganate
        Diatomaceous earth
        Slug killers based on aluminium sulphate

Prohibited   All other biocides including:
        Nicotine
        Formaldehyde
        Mercurial seed dressing
        Chemical soil sterilization
        Metaldehyde
        Jeyes fluid
        All other synthetic pesticides

Appendix C

Miscellaneous Standards

The following additions to the general production standards should be noted.

General

Restricted	Where fields are adjacent to land receiving herbicides, there should be an effective windbreak to prevent spray-drift contamination
	Pyrethrum, sulphur, diatomaceous earth in store — pre-harvest only, not on stored crop
	Rodenticides — only outside storage containers
Prohibited	Use of sprayers which are also used for prohibited substances
	Insecticides or fungicides, either on the harvested crop or in-store

Arable

Prohibited	Sprout inhibitors for potatoes
	Growth regulators
	Post-harvest burning of straw, cereal waste and stubble

Grassland

Permitted	Molasses as a silage additive
Restricted	Bacterial silage additives — Symbol approved only
Prohibited	Other silage additives

Livestock (Dairy)

Restricted	MAFF approved sterilants for milking parlours and dairies — only hot water sterilants allowed.
Prohibited	Penicillinase

190

Horticultural Systems

A lack of technical development and limitations of current farm structures have necessitated a number of special allowances for growers to be made.

Permitted   Loam and peat based seed, blocking and module composts using Symbol approved proprietary organic fertilizers for nutrient supply
Peat
Vermiculite
Perlite
Sand
Untreated bark products

Restricted   Introduced bulk organic materials from non-farm sources — only with Symbol Committee approval. These must be composted.
Conventional bare root transplants allowed until 1 January, 1989
Conventional modules and block plants allowed until 1 January, 1989
Conventional proprietary seed/blocking composts allowed until 1 January, 1990

Appendix D

Microbiological perspectives in soil management

E. John Wibberley, Royal Agricultural College, Cirencester

Soil fertility was once conceived in terms of earthworm activity with resultant effects on pasture productivity and potential liveweight gains of grazing stock. Such concepts came to suffer accusations of imprecision, of 'muck and mystery'.

From Boussingault's classic trials to Lawes and Gilbert founding Rothamsted in 1841, soil fertility became increasingly perceived in chemical terms. Universities founded departments of soil chemistry, only to be designated soil science departments well into the present century.

Soil management is an even newer terminology associated as it is with the improvement of soil physical properties. Physics came to take over the limelight from chemistry with the fuller appreciation of the significance of soil structure, embodied for instance in the Strutt report of 1972, *Modern Farming and the Soil* — which also stressed some concerns about chemical excesses on the land such as high usage of nitrogenous fertilizers leading to excessive lime displacement. More recently emphasis has been on compaction and its avoidance, subsoil management (so ably expounded by Professor Gordon Spoor and his colleagues) and, more latterly still, on soil

Extracts from article first published in *Soil and Water*, vol. 12, no. 2 (October 1985).

erosion in the UK. All this has set the seal on the importance of soil physical management.

What then of soil biology? How does it rank as a valid claimant to undergird practical soil management strategies?

Soil fertility criteria

Soil fertility has been perceived as a long-term quality of the soil in Britain — 'the original and indestructible powers of the soil' as stated by Ricardo in his classic rent theory. Doubt has been cast on the indestructibility of some British soils already.

Soil fertility refers to the total capacity of a soil to produce and go on producing useful crop yields. It is an all-embracing concept, leading logically to a search for sustainable management strategies.

A fertile soil supplies a crop with all its current-account needs now, viz. space, anchorage, water, air, warmth, favourable pH, nutrients and freedom from poisons and restrictions. It needs also to be conserved to supply these to future generations of crops. It needs to be treated like a deposit account in a dependable bank — yielding long-term rewards with interest.

Soil is both the beginning and the ending of agricultural life, in which the role of micro-organisms is foundational (figure 1). Soil

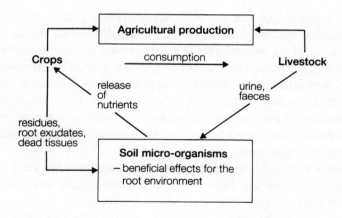

Figure 1   Soil micro-organisms: fundamental agents in agricultural production

improvement is both the lifeblood and the necessary consequence of any system worthy of the description 'sustainable agriculture'.

It seems obvious that the best likely indicators of a soil's suitability now to support a crop are micro-organisms requiring similar conditions for life to those favouring roots whilst their relative decline is the most appropriate portent of future deterioration in cropping potential.

Soil is a living system, diverse and intensely competitive at its best. It is possible to detect physical perfection in an environment whilst chemical faults may persist; such a state would result in biological poverty. It is possible to detect chemical Utopia coincident with physical unsuitability of environment. However, biological activity is only optimized when *both* physical and chemical environments are truly ideal also. Thus a plea is made for a micro-biological perspective as the paramount indicator of soil fertility once more — but with a difference by contrast with last century, viz. that research must supply evidence and facts to use as yardsticks in practical soil management with a microbial perspective. We must come in farming to value microbes more highly and to understand their *diversity* and their *activities* more thoroughly.

Diversity of the soil population

The *microflora* includes bacteria, actinomycetes, fungi and algae. The smallest and most numerous members of most soil populations are *bacteria* which are unicellular plants around one-thousandth of a millimetre in size. A saltspoonful of soil can contain as many as the human population of the world i.e. approaching 5000 million. Some can produce great-grandchildren within the hour. Far less numerous but also unicellular are the threadlike *actinomycetes* some of which are accredited with producing the characteristic splendid aroma of freshly-cultivated earth.

Fungi often tackle the more tough crop residues returned to soils and being multicellular, threadlike and fragmented are more difficult to enumerate individually so microbiologists often use the term 'biomass' to convey total weight of living matter rather than numbers. Data of the present writer concur with the work of others in showing a decline in the presence of fungi relative to bacteria as soil pH rises. . . . Most soil fungi are smaller than above-ground relatives and commonly exist as around 1 mm long fragments though many occur also as spores, some being very prolific in this respect.

Surface soils contain various types of *algae* including nitrogen-fixing blue-greens and silica-coated diatoms of diverse shape. The soil *fauna* ranges from such large creatures as earthworms, potworms, and springtail insects through eelworms (nematodes) of around 1 mm in length to the *protozoa* or microfauna represented by the well-known *amoeba* some ten times the size of the average bacterium.

Activities of the soil organisms

Beneficial effects

Decomposition. The largest-scale operation undertaken by the majority of soil micro-organisms is to break down crop residues, faeces and other raw organic matter (OM) to form humus (virtually fully decomposed OM of very high nutrient and water-retaining capacity and a valuable soil adhesive). In the process potentially harmful wastes which might otherwise be harbouring pests, pathogens and poisons are removed, mineral elements are liberated and humus reserves replenished.

The rate at which a soil can decompose raw OM such as crop residues is a function of the composition and toughness of the material involved, the size and diversity of the soil population and its level of activity as governed notably by soil temperature, pH, moisture supply and aeration. In a farm business it is not the amount of capital that is tied up which is so relevant to profitability as the rate of *turnover* of that capital. So it is with soil OM — I believe we should seek to measure *OM quality* and the influence of soil management practices on its *turnover* rather than absolute amount present at any one time. . . .

Mineralization. This involves the release of mineral nutrients from both OM and rock-particle sources by the action of the enzymes and

195

other substances produced by microbes. These minerals are then available for use by the current generation of crop roots.

Special chemical transformations. Some microbes carry out very specific chemical changes, for instance *nitrification.* This involves two genera of bacterium, *Nitrosomonas* and *Nitrobacter* which respectively convert ammonium to nitrite and nitrite to nitrate. Since nitrate is more soluble than ammonium it is more available for root uptake but also more easily leached. Therefore, there is current interest in bacteriostats such as 'didin' which limit the activity of these bacteria so that more nitrogen remains in ammonium reserves in soil and is more steadily released. . . .

Nitrogen-fixation. Apart from the well known *rhizobia* which inhabit legume root nodules there are various free-living species of bacteria and blue-green algae in soil which contribute useful amounts of nitrogen from air to soil.

Soil aggregation. Many bacteria are coated by gums which act as natural soil adhesives; fungal wefts deposit humus as they die and decay *in situ*; earthworms incorporate OM and concentrate calcium in their casts (2.5 million worms to the hectare can raise over 80 tonnes of casts in a year). All these processes promote the development of a more stable crumb structure in soil. . . .

Antibiotic production. Naturally produced substances which, in small quantities, inhibit the growth of other organisms originate in many microbe bodies. Certain actinomycetes such as *Streptomyces* are commercially cultured to collect antibiotics. These antibiotics in soil are important in defining territories for their originators. The producers are often decomposers which need to colonize a defined area of OM or of living crop root surface (rhizo-plane) — the immediate root zone is the rhizosphere and high microbial activity here results from a supply of root exudates (organic nutrient compounds) and the microbes in turn release minerals for roots. They also produce antibiotics which resist the penetration of the rhizosphere by pathogenic root-infecting fungi, for instance. Biological control can occur in other ways — for instance certain fungi can trap and kill nematode pests. . . .

Mycorrhizal symbiosis.   This occurs when certain fungi coat the root system of a crop so benefiting their own lifestyle but also effectively extending the crop's root system and therefore its recovery of less accessible nutrients such as phosphates. . . .

Detoxification.   Pollutants arrive in soils through industrialization, accident and deliberate addition for crop protection purposes. We depend on micro-organisms to render all these ultimately harmless by decomposing them. Investigations of their capacity or otherwise to do this must accompany developments in types of material added to soils.

Possible detrimental effects

Putrefaction.   Producing toxins in badly-aerated, cold soils overloaded with OM.

Denitrification.   Leading to loss of nitrogen as gases from very wet soils.

Immobilization.   The competition for very limited supplies of available nutrient which temporarily locks it up in microbial tissue instead of leaving it available to crop roots.

Pathogenicity.   Some microbes are agents of crop disease themselves.

All of these possible detrimental effects can be clearly linked to defective soil management. A soil physical and chemical environment favourable to crop roots also favours beneficial microbial activity.

A long-term appreciation of the roles of soil micro-organisms is needed for effective farming and as farmers we must call for and support continued relevant research in this relatively neglected field.

Appendix E

A proposal for a joint sheep enterprise at Rushall, June 1972

1 Introduction

The owner, Mr C. B. Wookey, farms some 5,000 acres at Upavon and Rushall, near Pewsey, Wiltshire. The Rushall farm, to which this proposal relates, consists of 1650 acres of owned land together with some 395 acres of downland rented from the Ministry of Defence. The owner is interested in chemical-free farming and hopes one day to be able to farm Rushall without recourse to artificial fertilizers or sprays.

To this end, it is proposed to reintroduce sheep to the farm to help with the fertility. As the owner's commitments are already heavy it has been decided to introduce the sheep as an entirely independent financial enterprise, with the owner making the land available and producing the necessary crops and the applicant providing the sheep and all related equipment and the labour.

The farm is farmed by a manager under the general direction of the owner, and great emphasis is placed on tidiness and good husbandry. The applicant would have to work in with the manager and be prepared to accept the standards already being observed.

2 Present policy

A three-year ley system of farming is in operation, with some 800 acres of corn, 450 acres of leys and 120 acres of various roots. In

198

addition, there are about 300 acres of permanent pasture on the lower land and 395 acres of downland. A beef enterprise utilises the grass, Friesian × Hereford steers being sold fat at from 10 to 12 cwt. The total varies from 600– 800 at any one time.

3 The requirement

In order to increase the fertility of the farm it is felt that the equivalent of one lambing flock of 500 ewes is necessary. Other enterprises would be considered provided the applicant could show that they would be viable and able to make sufficient contribution to the overall fertility.

4 Capital

The capital required for the enterprise would be found by the applicant. This would include, apart from the sheep themselves, all the necessary equipment, such as electric netting, mobile handling pens, spray race, hay cages, etc. In addition, a light tractor would be necessary to move the above and to haul feeding stuffs in the winter.

In the event of the applicant having insufficient capital to launch the enterprise and to live until the first return is to hand, the owner is prepared to provide up to 25 per cent of the capital required for the sheep only, at 10 per cent per annum to be a first charge on the profits.

5 Land

It is proposed to divide the year into two parts for the purposes of land usage. From 1 April to 31 July the sheep would be kept on an agreed acreage which would be the sole responsibility of the applicant. From 1 August to 31 March they would act as scavengers round the rest of the farm at the discretion of the owner, who would undertake the provision of maintenance for the sheep for these months. During the winter months hay would be provided together with clean swedes (to be hauled by the applicant) to provide maintenance. (Allowance should be made in the budget for severe weather when the swedes may be frosted.) None of the existing fences is sheep-proof.

6 Keep

The keep outlined above will be provided at the rate of 7½p per week per sheep. If lambing is to be part of the enterprise, lambs will be charged at 2½p per week per lamb from a date six weeks after lambing starts. Any supplemental feeding would be provided by the applicant.

7 Policy

The policy of the sheep enterprise would be discussed and agreed jointly by the owner and the applicant, but the owner would retain the right of overall control regarding which fields he wanted them to go onto at any particular time. In particular, the sheep would be expected to graze the undersown leys after harvest, and to run over the parkland areas when desired to do so by the owner.

8 House

There is a good semi-detached thatched cottage available to the applicant in the village of Rushall. This would be let unfurnished to the applicant at £4 per week plus rates and subject to Section 16 of the Rent Act 1965.

9 General

The sheep to be the sole responsibility of the applicant at all times and the owner will only be responsible for the provision of the agreed acreage of grass with one waterpoint per field. The grass will receive 3 cwt of a complete 1:1:1½ type fertilizer in the spring of each year. Any additional fertilizer would be at the applicant's expense. The applicant will be expected to keep well grazed (and topped if necessary) the agreed acreage for the April— July period. The applicant should be able to augment his income from the sheep by working on the farm at busy times by arrangement with the Manager at craftsman's rate of pay.

10 Application procedure

On receipt of this proposal, any applicant who wishes to be considered should prepare a full detailed scheme to be sent to the owner. This scheme should include the following points.

1 Type of sheep enterprise envisaged.
2 Numbers of sheep and breed.
3 Type of management.
4 Acreage of grassland required for the April— July period.
5 Finance:

   (a) Capital available.
   (b) Capital required from owner.
   (c) Budget.
   (d) Anticipated profits.

The applicant is also requested to append a brief outline of his personal history — age, education, present status, previous experience, etc. A note as to his ultimate aspirations would also be appreciated.

11 Conclusion

It is proposed to start the enterprise in the autumn of this year, 1972. Before submitting their detailed scheme to the owner, applicants can arrange to look over the farm by prior appointment with the manager. All applicants should have submitted their schemes by 15 July 1972.

A farming glossary

Acre	Measure of land area: 4,840 square yards; 0.405 hectare. There are 640 acres to the square mile.
ADAS	Agricultural Development and Advisory Service, operated by the Ministry of Agriculture.
Aftermath	The regrowth of grass after being cut for hay or silage.
Beat	Area of land looked after by a gamekeeper.
Binder	Machine to cut and tie corn into bundles (sheaves).
Blackthorn winter	A cold spell in May that coincides with the flowering of the blackthorn.
Calf group	An organization for collecting calves direct from farms, and selling to rearers without the calves' having to go through a market.
Carter	A farm worker who used to drive a team of horses.
Charlock	A yellow flowering weed (*Sinapis arveneis*) common in corn on the chalk before chemical sprays were introduced.
Cleft oak piles	Fencing posts made by splitting lengths of oak (as opposed to sawing them).
Comber (or reeder)	A machine to remove the grain from sheaves of wheat by combing first one end then the other to leave a bundle of straight, unbroken straw for use as thatch.

Compounding charge	The increase in price when straight fertilizers are mixed to produce a fertilizer with the required plant-food proportions.
Corn	In Britain, the generic name for wheat, oats, barley and rye. In America these are called 'small grains', and 'corn' refers to maize only.
Dayman	The old-fashioned general farm worker, who performed mostly manual tasks. A term no longer in use, as nearly all farm workers now drive tractors.
Debeaking	The clipping of the top mandible of a bird's beak to prevent feather-pecking.
Dipping	The statutory procedure of immersing sheep in a bath of insecticide to control scab.
Direct reseed	The process for establishing a ley without a nurse or cover crop. The seeds are broadcast or drilled onto bare ground and rolled, either in spring or in September.
Drenching	The administration of a dose of medicine to animals. Often done with a semi-automatic drenching gun that dispenses the required dose when a handle is pressed.
Dressed seed	Seed that has been treated with a chemical to control pest or disease.
Drilling	Planting, either corn, grass or root seeds, by a machine that places the seeds in rows at the required depth in the ground.
Experimental husbandry farm	Farms run by the Ministry of Agriculture in various parts of the country to try out new practices and/or crops, and to demonstrate farming practices.
Export restitution	The money paid to exporters by the Government or EEC under the Common Agricultural Policy to enable them to compete on the world market.

Fallow, bare	A field left ploughed and unplanted.
Fallow, bastard	A field that is ploughed after hay has been taken and fallowed for the rest of the summer to destroy the turf and any weeds.
Fallow, summer	A field left uncropped for the whole summer and cultivated continually to destroy all weeds.
Folding (sheep)	A system of giving sheep a daily ration of food by moving their pen-fold each day.
Growing out	Once corn, particularly wheat, is ripe and then there is a lot of damp warm weather, the grains will start into growth while still standing in the ear. This renders the corn unsuitable for seed or flour.
Growth inhibitor	Chemical used to prevent sprouting in stored potatoes.
Growth regulator	Chemical used to prevent corn from growing too tall, which leads to it going flat ('lodging').
Hagberg Test	To measure the suitability of wheat for bread-making. Based on the time taken for a given weight to fall through a paste made from the wheat under test. The longer the falling time, the better.
Hard wheat	A term used to denote a variety generally considered to give a milling sample. Protein content 11 per cent or above. Hagberg falling number about 250.
Harrow, seed	An implement consisting of a horizontal frame with iron spikes set pointing downwards. Drawn through the ground by a tractor, it gives a light raking effect.
Harrow, drag	Similar to the seed harrow but with a heavier and bigger frame and longer spikes.
Harrow, disc	An implement with two banks of metal discs set at a slight angle to cut and break ground, especially useful when turf has

been ploughed, to produce a seed-bed.

Hectare — Measure of land area: 2.47 acres.

Herbicide — Chemical to destroy growing plants. Some, such as Round Up, destroy all green growth. Others, called selective herbicides, destroy weeds in corn while leaving the corn unaffected.

Intervention — The arrangement in the EEC under the Common Agricultural Policy whereby governments buy up a commodity when the market price falls below a pre-determined level. This is then stored in intervention stores until released to be sold on the world market with the help of export restitutions.

Ley — A field of temporary grass. A long ley is usually for three to five years, after which the field is ploughed for wheat. A temporary ley is for one or two years. A ley mixture is a seed mixture for establishing a ley.

Mineralization — The conversion of the nitrogen in organic material into plant available nitrate.

Outliers — Cattle living outside in fields, not in yards.

Plough, chisel — Really a heavy-duty cultivator. A strongly made frame with six to eight tines that will break up hard ground. Needs a strong tractor.

Plough, mouldboard — The traditional plough that inverts the furrow as it is cut.

p.p.m. — Parts per million. Measurement used for trace elements, etc.

Reed — Wheat straw that has been combed to make thatch.

Roots — Generic name for fodder crops — kale, swedes, turnips, fodder beet, mangolds.

Scab — A mite induced irritation on the skin of sheep that causes loss of fleece.

Glossary

Shedding	Loss of grain from over-ripe corn. Once ripe, strong winds, rain or hail will cause grain to fall out of the ear.
Spring wheat, barley, etc.	Varieties of corn sown in the spring. Less hardy than winter varieties and so unsuitable for autumn sowing.
Spectacles	A device used to restrict vision on poultry to prevent feather-pecking.
Strong stores	Young (18—24 months) cattle with good frame ready to be fattened.
Stubbles	The short stalks of corn left after being cut by combine or binder.
Teaser ram	An impotent ram run with a flock of ewes to encourage them to come into season.
Thistle spud	A small blade affixed to the end of a stick to cut thistles below ground level.
Tilth	The fine seed bed required for planting.
Top-dressing	The practice of spreading fertilizer on the surface. The rain washes it into the ground for the plants to utilize.
Tramlines	The creation of unplanted wheelmarks in cornfields to facilitate top-dressing and spraying. Designed to fit the width of fertilizer spreader and sprayer so that no misses occur and to minimize damage to the growing crop. Not seen on an organic farm!
Tupping	Putting the ewes to the ram.
Undersowing	The practice of broadcasting (or drilling) grass seeds into a growing crop of corn. The development of the grass seeds is retarded by the height of the corn but after harvest they grow as they have more light and a grass field develops.
Volunteer cereals	Seeds from a previous crop that grow in the new one, e.g. wheat growing in winter barley.
Waney edge	A method of sawing planks from elm trees

	leaving one edge uncut, giving the 'waney' or 'wavy' effect. Used for cladding buildings.
Warble fly	A fly (*Hypoderma bovis*) that lays its eggs on the legs of cattle. As they hatch, they burrow their way through the body until they arrive under the skin of the back, where they develop into large larvae with an air hole in the hide. When fully grown they emerge, fall to the ground, pupate, and the cycle starts again. The whole cycle takes one year.
Winter wheat, oats, etc.	Hardy varieties of corn sown in the autumn.
White-string stage	When seeds germinate, they first send down roots, then produce the stem. Before this reaches the surface and light it is a white string.
Withy	Local name for willow.
Wood, hard	Beech and oak are examples of the hardwoods grown at Rushall. Slow-growing, maturing at about 100 years for beech and 150 for oak.
Wood, soft	Mostly the conifers — spruce, Scots pine and larch. Used on the estate for fencing.
Working down	The sequence of cultivations needed on a field to prepare a seed bed: plough, cultivate, drag, harrow, roll.

Index